RISK ASSESSMENT

RISK ASSESSMENT

Principles and Applications for Hazardous Waste and Related Sites

by

Peter K. LaGoy

np

NOYES PUBLICATIONS
Park Ridge, New Jersey, U.S.A.

Library of Congress Catalog Card Number:94–2510
ISBN: 0–8155–1349–6
Printed in the United States

Published in the United States of America by
Noyes Publications
Mill Road, Park Ridge, New Jersey 07656

10 9 8 7 6 5 4 3 2 1

Library of Congress Cataloging–in–Publication Data

LaGoy, Peter K.
Risk assessment : principles and applications for hazardous waste and related sites / by Peter K. LaGoy.
p. cm.
Includes bibliographical references and index.
ISBN 0–8155–1349–6
1. Hazardous wastes--Risk assessment. 2. Hazardous waste sites–
TD1050.R57L34 1994
363.72'876--dc20 94–2510
CIP

About the Author

Mr. LaGoy is Risk Assessment Manager for OHM Remediation Services Corp. and is located in their Hopkinton, Massachusetts office. Prior to joining OHM, he worked for Environmental Strategies Corp. in San Jose, California and ICF-Clement Associates in Washington, D.C. Mr. LaGoy has over 12 years of experience in the areas of site-specific risk assessment, cleanup criteria development, and regulatory toxicology. He has prepared over 50 site-specific risk assessments, exposure assessments, and quantitative toxicological assessments and has critically evaluated the effects of toxic chemicals and ionizing radiation on human health and the environment. Mr. LaGoy has B.S. and M.S. degrees in Biology from Georgetown University. He is a member of the Society for Risk Analysis and is the author of over a dozen papers on risk assessment.

Preface

> "The overarching mandate of the Superfund program [and of all remediation work, ed. note] is to protect human health and the environment from current and potential threats posed by uncontrolled releases of hazardous substances (1). "

Risk assessment can be thought of as the process used to measure the need for, and success of, remedial action.

Risk assessment at hazardous waste sites involves the use of standard approaches and assumptions in a reasoned, common sense manner. The purpose of this book is to provide practical guidance to people wishing to learn about, conduct, or use risk assessment procedures in evaluating hazardous waste sites. Critical issues, standard formulas and assumptions, and guidance on characterizing risk results in a useable manner are presented. The use of risk assessment as a key tool in selecting appropriate remedial options at hazardous waste sites is also described. Most attention is given to human health risks associated with exposure to toxic chemicals, but descriptions of the strategies used to estimate radiation health risks and to evaluate risks to the environment are also provided. Although most commonly applied to hazardous waste site remediation, the procedures outlined in this book are generally applicable to any situation involving a potential for health risks to an exposed population.

This book provides guidance on the mechanics of risk assessment preparation and illustrates these approaches with examples. However, the focus of the book is on the subjective nature of risk assessment, the art rather than the science. My brother's archery instructor in elementary school used a label maker to put a note on my brother's bow that was visible each time he shot. The note said "THINK". In risk assessment, as in archery, the mechanics may be perfect but the results off target unless the whole picture is integrated.

The actual risk (i.e., the right answer) can never be known. Consequently, while it is valuable to attempt to obtain the best numeric solution, reasonableness and the application of common sense are equally important. The book therefore devotes a substantial amount of space to issues of uncertainty that are inherent in risk assessment, and the need to address this uncertainty.

Preparing a book is a substantial task. I would like to acknowledge the support and assistance provided by my wife Patty during this process. Alice, age one, helped in her own way. Colleagues and friends who assisted by commenting on chapters include Tom Golojuch, Cris Jespersen, Steve McGinn, Jim Park, Mary Paxton, Tammy Quirk, and Beth Silverstein. Diane Bridgwater prepared the figures for the two case studies. Their assistance is greatly appreciated.

Table of Contents

1.

INTRODUCTION

Site-specific risk assessment is the process of evaluating whether or not a site poses a risk to human health or the environment. The purpose of all hazardous waste site remediation is ultimately to render a site safe for human or ecological populations. Consequently, risk assessment, as the process used to measure the effectiveness of the remediation process, is critical to all hazardous waste-site work.

1.0 PURPOSE AND PRINCIPLES OF SITE-SPECIFIC RISK ASSESSMENT

Risk assessments for Superfund have been defined as having two main functions: 1) To establish whether or not the Superfund site poses an imminent and substantial endangerment to public health, welfare, or the environment, and 2) to serve as a basis for establishing cleanup criteria (2). The risk assessment process is also used in comparing different potential remedial alternatives to ensure that the cure is not as bad as or worse than the initial problem. The current approach to risk assessment as outlined in EPA's Risk Assessment Guidance for Superfund (RAGS) focuses on an elaborate determination of the imminent and substantial endangerment (3). However, it is less useful in providing answers on appropriate cleanup goals. Certain areas in the current risk assessment process can be streamlined to increase the efficiency of preparing the assessment and to increase the usefulness of the final product.

Superfund risk assessments designed to force cleanup focus on maximum potential use scenarios and exposure assumptions for these scenarios. However, these risk assessments are not particularly useful in defining appropriate remediation goals or in planning appropriate remedial action. A risk assessment specifically designed to focus on the appropriate remedial action/cleanup goals for a site does not need to focus strictly on maximum use scenarios but can be tailored to likely possible uses. A consideration of likely site risks is critical in making truly informed decisions at a hazardous waste site. Risk assessment can be used as one element of an iterative process of evaluating possible site uses and remedial options in order to select the most appropriate option based on consideration of desired use, technical feasibility, costs, and risk. It may well be that the future of risk assessment lies not with the detailed, elaborate, and often ignored baseline risk evaluation but with its use as a tool in designing and implementing appropriate action.

Ideally, risk assessment would be a strictly objective process with decisions on risk based solely on scientific evidence. The National Academy of Sciences (4) noted that risk assessment (the science) should be kept separate from risk management (the policy). However, a considerable amount of uncertainty is inherent in the risk assessment process because of natural human variability in sensitivity and behavior, uncertainty in our knowledge of the potential for human exposure, and a paucity of information on the toxicity of chemicals to humans. Consequently, subjective value judgements have been a part of risk assessment from its inception. For example, the assumption that humans drink 2 liters of water each day is a standard risk assessment value, even though this value probably overestimates water consumption for most individuals. A policy decision (to ensure the protection of public health by using conservative values) has been incorporated into the risk assessment process and is treated as if it were a strictly objective value. As the primary purpose of site-specific risk assessments is to protect human health and not to get the most scientifically correct answer, policy decisions are, of necessity, incorporated into the risk assessment. As risk assessment matures as a science, hopefully the separation of the science of risk assessment from the application of risk assessment will occur naturally.

The field of toxicology provides a good example of the type of separation between scientific and policy issues that is necessary and possible. Toxicology, the science, is designed to answer the question: What dose causes a given effect? Regulatory toxicology, which incorporates policy considerations, is designed to answer the related but different question: "What is a safe dose?", in order to ensure health protection in the face of uncertainty. It would be helpful to have a term such as regulatory risk assessment to refer to decisions that are based on ensuring health protection and an alternative term such as scientific risk assessment for scientific discussions. In this book, as in current risk assessment practice, the distinction between science and policy will occasionally be blurred.

It is important to remember that the term "risk assessment" is used to refer to both scientific and policy aspects of the process, and to not confuse the two areas when they should be separated. It is equally important to remember that site-specific risk assessment almost always contains elements of both science and policy. One of the most commonly heard statements from regulators is: "You do the risk assessment, we'll make the risk management decisions." This statement ignores the risk management decisions that are incorporated into the risk assessment approach. Risk management, the process of using the results of a risk assessment to make decisions, will not be addressed in detail in the book. However, explaining the policy basis for results and decisions is an important part of the risk assessor's job at a hazardous waste site remediation.

2.0 HISTORY OF SITE-SPECIFIC RISK ASSESSMENT

The process of site-specific risk assessment is a fairly new process but its roots are firmly based in science that has existed for thousands of years (5). One can say that the process of risk assessment is not new. What is new is the process of attempting to predict the potential for harm in the absence of any clear evidence that an effect has occurred or could occur. The attempt to predict very low risks has led to an elaborate scientific approach to more accurately quantify the potential for harm to be associated with a given site. In this regard, risk assessment is most indebted for its development to the regulatory decision by the newly formed EPA in the early 1970s to assume that no threshold exists for carcinogens and that consequently, even a single molecule of a carcinogen can pose some, albeit very small, health risk. Searching for a "safe level" of a particular chemical then became the scientific and regulatory Holy Grail. At approximately the same time that cancer risks were being evaluated, EPA and Congress were becoming aware of a major and growing concern on the part of the U.S. population about the hundreds of industrial dumps and other waste sites located across the country. Fueled by concerns at such places as Love Canal, Congress passed the Comprehensive Environmental Response, Compensation, and Liability Act (CERCLA; aka the Superfund Act) in 1980. This law required sites posing an imminent and substantial endangerment to human health, welfare, or the environment, to be remediated.

In the past ten years, huge advances have been made in the site-specific risk assessment process. In the late 1970s and early 1980s, risk assessments were prepared on a case-by-case basis with each individual or firm performing the assessments differently and often using different exposure assumptions and approaches. In 1986, the EPA established guidelines for risk assessment in the Superfund Public Health Evaluation Manual (6). These guidelines, subsequent versions such as the Risk Assessment Guidance for Superfund (RAGS), and various support documents have provided clear guidance on preparing regulatory risk assessments and have greatly increased their consistency (3,7,8,9,10).

In the 1980s, the primary emphasis of a Superfund risk assessment was to determine if a site posed an endangerment to public health or the environment. The risk assessment provided written justification for the Agency's contention that a site posed a risk to public health and consequently had to be remediated, and was used to force action by potentially responsible parties (PRPs). The need for this type of risk assessment was necessitated by an understandable unwillingness on the part of many PRPs to accept huge cleanup costs for sites they considered unlikely to actually pose health risks.

Through many court cases, it has become clear that the Superfund law has sufficient power to force PRPs to take action and pay for cleanup at sites. Accordingly, although most PRPs may still not agree that a particular site poses health risks, most are aware that the regulatory agencies can and will force them to take action to remediate their sites. Consequently, industrial companies have by and large re-focussed their efforts from arguing that no remediation is necessary to a cost containment strategy. The need for a detailed risk assessment document to support a demonstration that a site poses a risk has been minimized. Consequently, risk assessments are being used more and more to assist in developing remediation goals that are still health-protective but that allow some flexibility in selection of the remedial option.

3.0 GENERAL APPROACH

A site-specific risk assessment evaluates the potential threat to human health (Human Health Evaluation) and the environment (Environmental Assessment) posed by contamination present at a site. Preparation of a site-specific risk assessment consists of combining available site-specific information with standard risk assessment assumptions and formulas to estimate potential exposure, and then combining the exposure estimate with toxicity information to estimate risks. A key part of this risk assessment process is an evaluation of the uncertainties associated with the assessment process. The quality of the risk assessment is highly dependent on the quality of information gathered for the site. It is important for risk assessment staff to carefully review the available data to ensure that a sufficient data base for risk assessment is available.

An outline of a standard risk assessment is presented in Table 1. The first section of the risk assessment outlines the purpose and goals of the risk assessment and generally describes the approaches that will be used. Section 2 of the risk assessment should contain the results of the data review, and should provide background information on the site and characterize the nature and extent of contamination. The process used to select chemicals of potential concern should be discussed in this section. In Section 3, information is presented on the environmental fate and transport of the key site contaminants and on the key toxicological characteristics of these compounds. As part of the toxicological assessment section, health-based criteria, and other established criteria and guidance that may be helpful in assessing potential risks are identified. An exposure assessment, including a discussion of exposure pathways, a description of potential receptors, and a quantitative estimate of exposure, is generally presented in Section 4. Section 5 contains the risk characterization, in which information on the toxicity of the chemical (from Section 3) is combined with the results of the exposure assessment (from Section 4) to determine if the site poses a risk to public health or the environment. Applicable or Relevant and Appropriate Requirements

(ARARs; established federal and state criteria) and a discussion of the uncertainties in the risk assessment are an important part of this section. The conclusions are presented as the final section.

TABLE 1
Outline for a Standard Baseline Risk Assessment

1.0 Introduction

2.0 Site Characterization

- 2.1 Site Description
- 2.2 Nature and Extent of Contamination
- 2.3 Selection of Chemicals of Potential Concern

3.0 Characteristics of Key Chemicals

- 3.1 Environmental Fate and Transport
- 3.2 Toxicity/Hazard Evaluation

4.0 Exposure Assessment

- 4.1 Routes of Exposure
- 4.2 Quantitative Exposure Assessment
- 4.3 Environmental Exposure Assessment

5.0 Risk Characterization

- 5.1 Comparison with Criteria
- 5.2 Current Use Scenarios
- 5.3 Future Use Scenarios
- 5.4 Environmental Risk Evaluation
- 5.5 Risk Assessment Uncertainties

6.0 Conclusions

7.0 References

3.1 Review and Evaluate All Available Data

As an initial step in a site specific risk assessment, all available data should be evaluated and summarized to indicate potential exposure pathways, determine the frequency of detection of site chemicals, and determine representative and maximum concentrations of the chemicals. Data gaps should be identified and filled if possible. Information should be present on hydrogeology, geology, climate, demographics, and on chemicals detected in environmental media at the site.

Data are evaluated to determine the nature and extent of site contamination, to determine key site contaminants, and to identify any critical data gaps. Information on background chemical concentrations and on local flora and fauna are also collected. Critical issues to cover in an initial data evaluation include identification of the types and state of chemical materials present at the hazardous waste site, a preliminary determination of the adequacy of the sampling effort (e.g., Were soil borings and groundwater wells located properly? Are further samples needed to define the nature and extent of contamination?), and evaluation of the analytical specifications (e.g., were detection limits low enough to detect the chemicals of concern?).

A second part of the data evaluation phase involves conducting a site visit of the facility. The site visit should include an inspection of key site features and the surrounding areas. A key focus of a site visit is an estimation of the demographics of the area, in particular a review of the location of local schools and hospitals, and a general understanding of site factors that may influence the results of the assessment. For example, a river running through the site is not clear evidence that fishing will occur. However, if a site walk reveals waste monofilament fishing line, and other fishing paraphernalia carelessly but usefully (from a risk-scoping perspective) tossed aside, fishing and the ingestion of fish from the site should clearly be considered in the risk assessment. The site visit can also be used to develop an understanding of the type of habitat in the area and the types of flora and fauna that are present or likely to be present.

The value of a site visit cannot be overemphasized. The adage that a picture is worth a thousand words is clearly true in risk assessment. The information that can be gained is critical to a clear understanding of the site and in making qualitative decisions about the actual likelihood of risks at the site. It can also be important in preventing the risk assessor from looking rather silly. I once prepared a risk assessment for a site in Wisconsin contaminated with PCBs that, based on a review of site photos and maps, had a pond located at its center. Consequently, I elected to consider the potential for fishing to occur in this pond. Given the ability for PCBs to bioaccumulate, the fishing scenario and the potential consumption of PCB-laden fish was the driving exposure route in the risk assessment. Several drafts were prepared and the report was finally submitted in draft final form to be used with minor revisions as the basis for cleanup at the site. However, as the nature and extent of contamination was being more fully defined in anticipation of designing remedial strategies, we were asked to conduct a site visit. On arriving at the site, I was surprised to see that the "pond" that had been the basis of the risk assessment numbers, was essentially dry. When asked, the site investigators commented that they had wondered about the fishing scenario, but assumed that we were the experts and must know what we were doing. A site visit or even more thorough discussions with personnel familiar with the site would have prevented us from making an unnecessary error.

3.2 Select Key Contaminants (Chemical of Concern; Indicator Chemicals)

At many sites, numerous chemicals are detected in environmental media, including naturally-occurring chemicals (e.g., metals) as well as site-related contaminants. When a large number of chemicals is detected at a site, a subset of key chemicals of concern is often selected in order to focus the assessment on those contaminants that are most likely to pose risks at the site. The general criteria used to select the chemicals of concern are presented in Table 2.

The last two criteria listed in Table 2 are not commonly used in the selection process and there is some debate over whether or not to use a selection process at all. The selection of chemicals of concern was and occasionally still is one of the more contentious areas in a site-specific risk assessment. With desktop computing, it is often a rather simple process to carry all chemicals through the risk assessment. The results can be summarized in tables and only those chemicals contributing substantially to the risk at the site need be presented in the text. Carrying all chemicals through the assessment ensures that all compounds are adequately considered and also directly responds to any concerns that a reader of the risk assessment may have with regard to particular chemicals. Being able to show that cyanide, for example, does not pose a risk at the site may be more convincing to some audiences than arguing that it has rather low toxicity and is present at low levels relative to other site contaminants. The latter approach still leaves the question of "Isn't cyanide a health risk at the site?" unanswered, and still open to debate in the minds of audiences whose only knowledge of cyanide is its use as a poison.

Background. A continuing point of contention at many sites is the question of whether or not a chemical detected on site is present as a result of site contamination or is only present at background levels. Complicating this issue is the fact that background refers to both naturally occurring levels of materials (e.g., many metals) as well as to chemicals that have become ubiquitous in the environment as a result of human activities (e.g., DDT and PCBs). Even more interesting are those compounds, lead and the polycyclic aromatic hydrocarbons (PAHs) as good examples, that both are naturally occurring and derived from anthropogenic sources.

TABLE 2

Factors Considered in Selecting Key Site Chemicals

- Comparison with blanks. If chemicals are detected in travel or field blanks, concentrations of compounds in on-site samples and field blanks are compared. On-site concentrations of chemicals that are common laboratory or field contaminants (e.g. acetone, phthalate esters, methylene chloride, and toluene) that exceeded field blank concentrations by ten times (10x) and other chemicals that exceeded field blank concentrations by five times (5x) are considered to be site-related (3). Chemicals not meeting these criteria are treated as non-detects.

- Comparison with background concentrations. Concentrations of contaminants detected on the site are compared with concentrations detected in upgradient samples or in samples from off-site areas that are not contaminated by the site.

- Frequency of detection. When a sufficient number of samples have been collected to ascertain that the site has been well characterized with respect to any individual chemical, the frequency of detection for any individual chemical can be considered in selecting chemicals of concern. For example, chemicals detected in less than 5% of the samples may not be considered to be of concern at the site. However, dispersion patterns and the toxicity of chemicals eliminated in this manner must be reviewed to ensure that chemicals for which a hotspot of potential concern exists are not removed from consideration.

- Consideration of Concentration, Toxicity and Physicochemical Properties. The final selection of chemicals of concern may also include consideration of the magnitude of detected concentrations at the site, known toxic potential, and migration characteristics in environmental media. It may also include a consideration of potential breakdown products of the chemicals in the environment. For example, chemicals clearly not of toxicological concern may be removed from further consideration even if they satisfy certain of the other criteria outlined above. Conversely, compounds that do not meet one or more of the three selection criteria noted above, but which may have been used at the site and are known to be highly toxic may be selected. Finally, if high levels of a compound with a very toxic breakdown product are present, the breakdown product may need to be considered in the assessment even if it has not as yet been detected.

Regulatory agencies have fairly recently begun emphasizing statistical approaches to background sampling and comparison and the use of site-specific background data. The use of national or regional background levels published in the literature is discouraged. Previously, much simpler comparisons were used. For example, at one site, the background comparison consisted of comparing the average on-site concentrations with either 10 times or 100 times average background concentrations for the area (11). At another site, chemicals with on-site concentrations that were less than five times (5x) the background concentrations were considered not to be different from background. A factor of five was used to be consistent with the EPA-approved procedure for blanks noted above.

While the use of site-specific background data is generally valuable, there are times when other approaches, such as comparing on-site levels with literature values or even using a comparison with 10-times an average background level may be appropriate.

Reliance on rigid statistical methodology to arrive at background levels of contaminants may lead to erroneous decisions and inappropriate actions. For example, a site constructed on imported clean soil or conversely, on an area that was excavated, may have very different geology than nearby, and visually similar background sampling locations. In many cases, it may be easiest to carry chemicals that arguably are present at background levels through the risk assessment process. In this way extensive arguing about the appropriateness of the background sampling locations and the comparison techniques may be avoided. On the down side, some chemicals (arsenic is the classic example) pose risks at background levels and raising this issue may complicate discussions. Suffice it to say that the process of background comparison is often one of the more interesting steps in the risk assessment. The more information that can be collected and provided, the more likely that the correct solution will be found.

3.3 Evaluate the Hazard of the Key Chemicals

Information on the toxicity of the key contaminants should be obtained from appropriate peer-reviewed sources such as EPA's IRIS database (12), from other federal regulatory agency-approved sources, (many of which are listed in the EPA 1989 RAGS manual), or from appropriate State regulatory agencies. Sources of toxicity information that are based on occupational exposure should be used cautiously as they may not be appropriate for non-occupational exposures.

For carcinogens, the most commonly used toxicity criterion is the cancer slope factor (CSF, also called the cancer potency factor), a value that expresses the relationship between the probability of developing cancer and the dose (13). For example, a cancer potency factor of 0.1 per 1 mg/kg/day (written as 0.1 $(mg/kg/day)^{-1}$) indicates that continuous daily exposure at a dose of 1 mg/kg/day is associated with a 1 in 10 chance of developing cancer. Unit risk values, which express the risk in terms of a given unit, usually m^3 of air or liters of water, are also used but less frequently. Unit risk values function the same way as cancer slope factors, except that the exposure has been assumed and the values are expressed in terms of $(ug/m^3)^{-1}$ or $(ug/liter)^{-1}$. There is a good deal of uncertainty associated with developing cancer slope factors. In order to ensure that public health is protected, cancer slope factors are generally upperbound estimates of risk, that is, the actual risks associated with exposure to the chemical are unlikely to be higher than the estimated risks.

For noncarcinogenic or systemic effects (which can include the noncarcinogenic effects of carcinogens) the reference dose (RfD; formerly called the allowable daily intake or ADI) in units of mg/kg/day is used as the toxicity criteria (3). The RfD represents the dose of a given chemical that is considered unlikely to cause adverse effects in humans.

3.4 Determine Site-Specific Exposure Pathways

Based on the information available on the site and taking into account all potential future uses of the site, potential exposure pathways should be determined. The likelihood of exposure actually occurring via each pathway should be noted. For pathways that appear unlikely to occur or to lead to significant exposure, a detailed qualitative discussion of potential exposure may be used to drop the pathway from further consideration. For principal exposure pathways, a more in-depth evaluation of the available data to determine if they are adequate for quantitative risk assessment may be useful.

3.5 Quantify Exposure

For the principal exposure pathways, the measured concentrations of key contaminants in environmental media are used directly or incorporated into models to estimate exposure point concentrations, which are combined with various assumptions to calculate the dose of contaminant to an exposed individual. At most sites, the scenario used to assess maximum potential future exposure is on-site construction of a house with a private well (under no-action, this is a possibility at almost all sites and is likely to give the greatest risk), but other scenarios are also considered. Often, a reasonable maximum scenario (worst-case analysis) and an average scenario are assessed in order to determine the potential range of exposure.

3.6 Evaluate Risk

In the risk characterization section, hazard information is combined with exposure information to determine if contamination at the site poses a risk to public health. As a preliminary check, the derived risks are compared with target risk levels to determine whether or not chemicals at the site are likely to pose risks. For carcinogens, EPA and many other regulatory agencies typically use an upper-bound excess cancer risk level of 1 in 1,000,000 (10^{-6}) or 1 in 100,000 (10^{-5}) as a target risk goal. If risks are below this level using conservative (health protective) exposure scenarios and assumptions, then it is unlikely that the site poses a significant cancer risk. If estimated risks are above this level, it does not necessarily mean that the site poses unacceptable health risks, but that some further evaluation of the risk levels and possibly remediation may be required. The EPA generally regulates to within a risk range of 10^{-4} to 10^{-6}.

For noncarcinogens, the estimated exposure is compared with the appropriate toxicity criterion (usually the reference dose or RfD). If the estimated exposure using conservative assumptions is less than this criterion, then the chemical is considered unlikely to pose a risk. Again, if exposure levels exceed the RfD, it does not necessarily mean that unacceptable health risks are posed by the site, but rather that some additional study of the situation is necessary. At most sites, more than one chemical is present and exposure could occur by more than one route. The potential effects of multiple chemical and multi-route exposure need to be considered in the assessment. Finally, the uncertainties associated with the assessment must be considered in order to put the risk values in perspective.

4.0 SUMMARY OF REMAINING CHAPTERS

The overall format of the book follows the steps in a site-specific risk assessment (Chapters 1 - 5), with special topics covered in the remaining chapters. This introduction discusses the history and purpose of site specific health risk assessment in the hazardous waste site investigation and remediation process. It also contains a discussion of the key components of risk assessments. Chapter 2 describes the physical and chemical properties of hazardous substances that can influence their behavior in the environment and in the exposed organism. Factors that need to be considered in assessing the transport and fate of chemicals are also discussed. In Chapter 3, the toxicological information that is used in a health risk assessment is described. Standard approaches and assumptions are presented, along with the bases for these approaches/assumptions. The history of regulatory toxicology and expected changes are also noted. Chapter 4, Exposure Assessment, describes the process of selecting critical exposure pathways for quantitative evaluation, the standard approaches to quantifying exposure, and the basis and use of assumptions in exposure assessment. The final section of a risk assessment involves combining the results of the exposure assessment with the toxicity evaluation to characterize risks at the site. Chapter 5 describes this stage of the risk assessment. The chapter also includes a discussion of approaches that can be used to increase the useability and acceptability of the risk assessment results.

Two components critical to all risk assessments are a description of the uncertainty in the risk assessment and presentation of the results in a form that is understandable to its intended audience. Uncertainty is inherent in all site-specific risk assessments, a factor that is often ignored in establishing final remedial plans at a site. In Chapter 6, uncertainty in the risk assessment is addressed, the types of uncertainty that occur are described, and approaches for qualitatively and quantitatively evaluating uncertainty are outlined. Chapter 7 presents the key rules of risk communication, anecdotal information on do's and don'ts in explaining risk results, and aspects of human behavior that need to be considered in presenting risk results. Because public perception can have at least as important an effect on the outcome of risk assessment work as the science, and because scientists generally do not work well with the type of subjective issues involved in risk communication, consideration of communication/perception issues is critical.

The ultimate goal of all investigation and remediation work at hazardous waste sites is to ensure the protection of human health, the environment, and welfare. However, risk assessment results are often not considered or only considered tangentially in the final remedial decision. Chapter 8 describes a strategy for using risk assessment to assist in selecting the most appropriate remedial approach at a hazardous waste site. Options based on expected or desired future use of the site and risk-based versus technology-based remediation standards are considered.

In addition to chemical risks to humans, the effects of radionuclides and the effects of chemicals on the ecosystem are becoming an increasingly important part of hazardous waste site work. Chapter 9 summarizes terminology and generally serves as a primer for performing radiological risk assessment. Regulatory agencies have developed approaches to assessing risks associated with radionuclides at mixed waste sites that differ somewhat from the approaches commonly used to assess chemical-related risks. Chapter 10 describes approaches to assessing ecological risks through the use of both traditional risk assessment approaches as well as the use of field studies and bioassays. A discussion of information sources and the direction of this field is provided.

Two case studies are included in the book as Appendices. The first case study presents a standard risk assessment, including calculations, qualitative decision making, and the presentation of final results. The second case study presents an approach to developing risk-based remediation goals for a site. The interaction of risk assessment and the remedial process are briefly discussed.

5.0 GENERAL REFERENCES

Environmental Protection Agency (EPA), Risk Assessment Guidance for Superfund: Human Health Evaluation Manual, Part A, Interim Final. Office of Solid Waste and Emergency Response, Washington, D.C., OSWER Directive 9285.701A, July (1989)

Paustenbach, D.J., ed., *The Risk Assessment of Environmental Hazards: A Textbook of Case Studies.* John Wiley & Sons, New York (1989)

2.

CHEMICAL CHARACTERISTICS

The chemical and physical properties of compounds and elements present at a hazardous waste site have a profound effect on the level of risk posed by the chemicals. The potential for release, the mobility of the chemical in the environment, the distribution and metabolism of the chemical in the body, and even the toxic effects of the chemical, are all a function of its physicochemical properties. The behavior of chemicals in the body and their effects (i.e., toxicity) are discussed in Chapter 3. This chapter focuses on the impact of physicochemical properties on chemical transport and fate in the environment.

The transport and fate of chemicals in the environment are dependent on both the properties of the chemical and on the properties of the environmental medium in which the chemical occurs. Because of the effects of the latter, transport and fate of a compound can only be discussed in general terms when information on the characteristics of the environmental medium in which it is present and through which it may migrate are not available. In addition, the following presentation of general concepts of transport and fate given below is, by necessity, greatly simplified. Therefore, the principles discussed below can only be considered useful as general guidelines, and are not necessarily applicable in all cases. The general references listed at the end of this section provide a more thorough treatment of these issues.

1.0 CHEMICAL AND PHYSICAL PROPERTIES

The environmental fate and transport of contaminants is dependent on their physical and chemical properties, the environmental transformation processes affecting them, and the media through which they migrate. Physical and chemical properties are defined in Table 1. A brief discussion of the properties and their effect on the environmental behavior of chemicals is presented below.

The water solubility of a compound is a critical property affecting environmental fate: highly soluble chemicals can be rapidly leached from soils and are generally mobile in groundwater. Solubilities of organic chemicals range from less than 1 µg/liter to greater than 100,000 mg/liter (6). It should be noted that solubility is a relative term. Chemists consider most solvents, with solubilities in the range of 1,000 - 10,000 mg/liter, insoluble in water. However, hazardous waste professionals consider these same materials quite soluble in water. The difference is based on the point of reference, with chemists ranking solubilities based on a large universe of chemicals that include many that are totally miscible in water, as opposed to environmental scientists who typically deal with chemicals that have solubilities in a smaller range, generally from low µg/liter to 10,000 mg/liter levels.

The density of a compound indicates if it will have a tendency to float or sink either in surface water or in groundwater. Fluid movement through groundwater aquifers is influenced by density differences of as little as 1% (14). Chemicals less dense than water (floaters) will remain on top of the groundwater layer, chemicals with densities of approximately one (mixers) will move with the groundwater, and those compounds with a high specific gravity (sinkers) will move under the influence of gravity, possibly in a direction different from that of the groundwater. Transformation of the chemical in the environment can lead to a change in the specific gravity. Trichloroethylene, a sinker with a density of 1.5, may be biotransformed to vinyl chloride which has a specific gravity of 0.9 and is a floater (15).

TABLE 1

Chemical and Physical Properties

Atomic or Molecular Weight: The weight of an atom or molecule of a chemical expressed in atomic mass units. One atomic mass unit is equal to 1/12 of a carbon-12 atom.

Boiling Point: The temperature in degrees Celsius at which the vapor pressure of the compound is equal to or slightly greater than the atmospheric pressure (760 mm Hg).

Melting Point: The temperature in degrees Celsius at which a material changes state from a solid to a liquid at atmospheric pressure.

Specific Gravity (Density): The weight of a given volume of the compound at a specified temperature relative to the weight of an equal volume of water at 4 degrees Celsius.

Solubility in Water: The maximum amount of the chemical that will be totally dissolved in water at the specified temperature.

Solubility in Organics: The ability of the chemical to be dissolved in specified organic compounds.

Octanol/Water Partition Coefficient: The relative amount of chemical that can be totally dissolved in octanol compared to the amount that can be dissolved in water. Often expressed in a logarithmic scale as log octanol/water partition coefficient.

Vapor Pressure: The pressure in millimeters of mercury exerted by the vapor phase of the chemical at the specified temperature.

Vapor Density: The weight of a given volume of the chemical relative to the weight of the same volume of air.

Henry's Law Constant: An expression of the distribution of a chemical between air and water at equilibrium. Usually defined as the ratio of the partial pressure of the compound in air measured in atmospheres to the mole fraction of the compound in a water solution.

pKa: A measure of the extent of dissociation of a material, which is defined as the pH at which half of the compound is ionized. pH is defined as the logarithm of the reciprocal of the concentration of hydrogen ion in a solution.

Flash Point: The temperature at which a liquid or solid gives off enough vapor to allow ignition of the vapor/air mixture.

Volatilization of a compound will depend on its vapor pressure, water solubility, and molecular weight. Vapor pressure is a relative measure of the volatility of chemicals in their pure state and in general is a good predictor of potential volatility. Vapor pressures for organic liquids range from approximately 10^{-7} to 760 mm Hg (6). However, highly water-soluble compounds generally have lower volatilization rates from water than compounds having a low water solubility. For example, ethanol and 1,1-dichloroethylene have similar vapor pressures (740 mm Hg and 600 mm Hg, respectively) but very different volatilization rates. The Henry's Law constant, which combines vapor pressure with solubility and molecular weight, is more appropriate than vapor pressure alone for estimating releases to air from water. The Henry's Law constants are 4.5 x 10^{-5} atm-m^3/mol for ethanol and 3.4 x 10^{-2} atm-m^3/mol for 1,1-DCE, indicating that 1,1-DCE is much more volatile from water than ethanol. Compounds with Henry's Law constants in the range of 10^{-3} and larger can be expected to volatilize readily from water and soil. Those with values ranging from 10^{-3} to 10^{-5} volatilize less readily, while compounds with values less than 10^{-5} volatilize slowly (16).

The octanol-water partition coefficient (K_{ow}) is often used to estimate the extent to which an organic chemical will partition from water into lipophilic tissues of organisms, for example, animal fat. Similarly, the organic carbon partition coefficient (K_{oc}) reflects the propensity of a compound to adsorb to the organic matter found in soil. The K_{oc} can be calculated from the K_{ow} using any of several equations, with the most general being $K_{oc} = 0.544 \log K_{ow} + 1.377$ (10). The organic carbon partition coefficient may be a better indicator of chemical movement than water solubility for hydrophobic chemicals in soils with a high organic content, since the relative adsorption to organic materials may exert the greatest control on chemical transport in such soils. The normal range of

K_{ow} values is from 1 to 10^7, with higher values indicating greater adsorption potential. Data on total organic carbon (TOC) content is often not collected from hazardous waste sites but, because of the potential for organics to bind to soils, this parameter can provide useful information in evaluating the potential for migration to occur at the site.

2.0 TRANSPORT IN THE ENVIRONMENT

Chemicals at a hazardous waste facility can be transported off site by several pathways. The chemicals can be carried off by the wind after either volatilizing into the air or become airborne through the action of wind on surface soils. Chemicals can also leach downward through the soil and be transported off site in groundwater, or move downward along the bedrock as a dense nonaqueous phase liquid (DNAPL). Once in the groundwater, contaminants may discharge to surface water bodies. Finally, compounds can be carried off site and into nearby bodies of water by surface runoff from the waste site. There are other methods of transport, such as the tracking off site of contaminated soil by humans, animals, or vehicles, and the transport of contaminated dust by the wind. These pathways may be important in some cases, but in general, volatilization, atmospheric transport, groundwater transport, and movement in surface water are the main transport processes.

Movement of a chemical through the soil is dependent on the relative partitioning of the material between soil, water, and air, the density of the compound, the groundwater flow, characteristics of site soils, and other site-specific characteristics. Another important factor is whether or not the material is broken down by soil microorganisms. The general chemical and physical properties important for defining the transport of a compound through the soil are vapor pressure, water solubility, organic carbon partition coefficient and density (specific gravity). The potential for chemicals to migrate is described in the following subsections.

2.1 Percolation into Groundwater.

Contaminants present in soils at a site may leach to groundwater. Site-specific factors influencing the potential for downward migration to groundwater include depth to groundwater, soil characteristics (i.e., clayey or sandy soils), factors affecting adsorption such as TOC content of the soil (primarily affecting organics) and mineralogy (primarily affecting metals), the amount of rainfall and evaporation and the amount of runoff, which is dependent on the amount and type of cover material on the site. Depth to groundwater coupled with the types of soils any percolating water or chemicals have to migrate through will obviously influence whether or not groundwater is reached. The amount of binding that occurs as the chemicals move downward through the soil also affects the potential for the chemical to reach the water table. Little information on the adsorption potential of site soils is collected as part of most remedial investigations, and increased attention to collection of data in this area may be particularly useful in modelling efforts, risk assessment, and in designing treatment options.

Infiltrating precipitation is generally the key carrier for leaching contaminants from soil and transporting them to the groundwater. Clearly, the amount of rainfall and sunshine (evaporation) will substantially affect the potential for infiltration at a site. Finally, if much of the ground surface at a site is paved or covered with buildings, only small areas of the ground surface may be available for precipitation to recharge the groundwater. Similarly, if an effective grass or clay cap is present and properly sloped, chemicals may persist in vadose zone (unsaturated) soils for a very long time.

As noted above, the characteristics of the individual chemicals can greatly affect migration downward through soils. In addition, chemicals present in mixtures can influence the behavior of each other. Organic chemicals with relatively low organic carbon partition coefficients (K_{oc}s) are not likely to adsorb to soil organic matter. Most of these chemicals have solubilities in the hundreds-of-milligrams-per-liter range or greater and are expected to be fairly mobile in the aqueous phase (i.e. once they become dissolved chemicals in the groundwater). Organic chemicals that have a high tendency to adsorb to soil organic matter such as PCBs, organochlorine pesticides, and high molecular weight petroleum hydrocarbons are generally much less mobile. However, if these poorly mobile compounds are present together with more mobile constituents, their mobility may be enhanced by these compounds.

Concentrations of organic contaminants in groundwater that approach their aqueous solubility limits (e.g., TCE detected at a concentration of 860 mg/liter when its solubility is 1100 mg/liter) may indicate the presence of a non aqueous phase liquid (NAPL). NAPLs that are less dense than water (i.e., with specific gravities less than 1.0) are known as Light Non-Aqueous Phase Liquids (or LNAPLs) and NAPLs that are denser than water are known as Dense Non Aqueous Phase Liquids (DNAPLs). MacKay et al. (14) reported that separate non-aqueous organic liquid phases of a contaminant generally occur in groundwater when the observed concentrations of an organic chemical are greater than 10% of their solubility. Consequently, if levels of chemicals in groundwater are observed to be at or above 10% of their solubilities, the presence of a NAPL should be suspected.

The migration of inorganic chemicals through soil to groundwater is influenced by soil characteristics and water movement. Soil parameters that must be considered are clay and metal oxide content, fraction of organic matter, pH, and oxidation-reduction potential. Chemical partitioning between soil and water can be expressed by a soil-water distribution coefficient (K_d). K_ds may underestimate migration potential because they are based on chemical adsorption and desorption reaching equilibrium. As a rough estimate of relative mobilities, K_ds for inorganic chemicals can be compared (17). Cadmium has a low K_d of 6.5 whereas lead, with a K_d of 900, is generally much less mobile in the environment. However, mobility can be enhanced by the presence of other chemicals such as naturally occurring humic materials.

2.2 Migration into Air

Volatilization of a material from a hazardous waste site occurs in a number of stages: movement of the chemical from a source to a point of contact with air, transfer from the soil or water to the air, and finally movement with the air. A number of site-specific environmental factors such as temperature, precipitation, average windspeed, soil moisture content and soil pore space affect these individual processes. Certain properties of individual chemicals can also be important in this regard. The vapor density of a chemical influences whether it will move up or down in the unsaturated soil zone (the vadose zone) while the chemical's specific gravity will affect its behavior in saturated soils. Depending on the organic content of the soil, a chemical's partition coefficient may influence movement. Water-soluble compounds may be carried to the surface and evaporate with the water, but in moist soil volatilization is less likely. A chemical's Henry's Law constant is probably the most valuable indicator for determining the probability that a chemical will pass from the soil surface into the air, compounds with higher Henry's Law constants being most likely to "escape". Finally, the molecular weight of a compound is important in determining the rate of diffusion of the material through the layer of stagnant air above the soil surface. Therefore, information on the chemical and physical properties of a compound gives some indication of its relative volatility. For example, a poorly adsorbed, low molecular weight chemical with a high Henry's Law constant will evaporate more readily than a material with the opposite characteristics.

Groundwater from beneath a site may eventually discharge to surface water. Volatile chemicals entering surface water may be released into the air, depending on factors such as vapor pressure and solubility. Concentrations of contaminants released into surface water are likely to be fairly low due to attenuation by soils and the likely slow release into the water, and volatilization would probably result in very low and possibly non-detectable concentrations in the air, even immediately above the surface water.

Remedial activities can enhance the release of chemicals to air and these potential releases need to be considered in designing remedial systems. Contaminants adsorbed to soil particles may migrate into the air during excavation in the form of fugitive dust. Excavation can expose subsurface soils to the air and facilitate volatile emissions from this freshly exposed soil. Obviously, thermal destruction or thermal volatilization techniques can enhance the release of chemicals from a site. Even in situ remedial actions can lead to the release of chemicals. Air stripping of contaminated groundwater results in volatile contaminants being released. Emission controls such as carbon absorption are typically used to limit or prevent release of these compounds but the specific requirements for control vary from state to state.

2.3 Migration in Surface Runoff Water

Chemicals can move off site in surface runoff either adsorbed to particulate matter or dissolved in the water. Chemical properties influencing migration include solubility, octanol water partition coefficient, and density. Any chemical located near the surface can be transported in this matter. Once in a larger body of water, the particulate matter will tend to settle to the stream bottom along with any adsorbed chemicals, while volatile materials will generally evaporate. Other compounds will remain dissolved in the surface water until degraded, possibly being transported considerable distances.

3.0 FATE IN THE ENVIRONMENT

Chemicals in the environment will degrade or be deposited and stored based on properties of the medium in which they are found, the presence or absence of degrading microorganisms, climatic factors, and their intrinsic chemical and physical properties. The most common fate processes are the chemical processes of photolysis, oxidation, reduction, hydration, and hydrolysis; the biological processes of biodegradation, biotransformation, and bioaccumulation; and adsorption. Compounds are usually influenced by more than one process, with the relative importance of each process dependent on the environmental conditions. Photolysis involves the absorption of light energy by either the target chemical (direct photolysis) or by another substance which uses the energy to interact with the target chemical (indirect photolysis). The energy absorbed by the chemical causes it to be transformed into another compound. Oxidation of a chemical can occur under the influence of oxidants present in water due to photochemical reactions or from other sources. Under anaerobic conditions, chemical transformation due to reduction reactions may occur. Hydrolysis, the addition of a hydroxyl group to the chemical, is a common transformation process. It is promoted by either acidic or basic conditions depending on the chemical. Hydration is the addition of water molecules to a chemical.

Organisms present in the same environment as a toxic chemical can significantly affect its fate. Many microorganisms can use manmade hydrocarbons for food and thereby biotransform and biodegrade these compounds. The process of biodegradation like most biological processes is dependent on numerous factors (some currently known, others unknown) and estimating rates of degradation is not particularly exact at this time. Higher animals also can metabolize some toxicants, although, in general, microbial transformation is a quantitatively much more important process. Chemicals with high octanol-water partition coefficients may be stored by organisms. This bioaccumulation usually occurs in specific target tissues or in tissues with a high lipid content and the compound may also be concentrated in these tissues.

Materials that are adsorbed tightly to particles or to organic matter may be bound to certain types of soils and sediment. If bound tightly, this material may be effectively stored and thereby removed as an active constituent of the environment.

3.1 Transformation Processes in Groundwater

Transformation is dependent upon a variety of factors that may vary locally within a site. These factors include: availability of organic chemicals, availability of microorganisms, nutrients and growth factors, toxicity, and inhibition. For example, some halogenated aliphatics have been found to undergo reductive dechlorination under anaerobic conditions (15,18,19). On the other hand, trichloroethylene degrades aerobically in the presence of methane. The anaerobic transformation is sequential, with, for example, tetrachloroethylene yielding first trichloroethylene, then 1,2-dichloroethylene and, ultimately, vinyl chloride (15,19). 1,1,1-Trichloroethane, also, has been shown to undergo transformation to 1,1-dichloroethylene and then to vinyl chloride (15).

3.2 Cycling of Persistent Chemicals

Persistent chemicals will tend to cycle through the environment, moving from one medium to another. For example, chemicals that have evaporated or have been carried into the air on dust particles can be brought back to earth by precipitation or by random dry deposition and enter surface or groundwater. Groundwater often flows into surface water, carrying any potential contamination as well. Also, chemicals may move along different pathways based on the initial starting spot, i.e., on the surface or deep in the soil. Therefore, although a chemical will preferentially move into one medium or another, this does not preclude its appearance in a normally unpreferred medium.

Mercury serves as a good example of cycling in the environment. Inorganic mercury released into the air from sources such as coal fired power plants in the Midwest, travels to the Northeast and is ultimately scrubbed from the sky via precipitation. The mercury is deposited on surface soils, plant leaves, and in surface water bodies. Mercury in the leaves eventually is deposited on soils where it can be leached via surface runoff into streams and ponds. Aquatic organisms take up the inorganic mercury and convert some of it to organic mercuric compounds (methyl mercury). Larger predators eat the smaller aquatic organisms and the mercury moves to higher and higher trophic levels until it is re-deposited through the death of the top organisms to sediments (aquatic predators) or to the forest floor (avian or mammalian predators). Mercury on land can be leached back into the surface water through surface runoff. Mercury in the aquatic sediments will be taken up by benthic organisms, which will in turn be taken up by larger organisms and ultimately by top predators and recycled through the aquatic system.

3.3 Summary

In summary, the two broad categories for environmental fate processes are accumulation and degradation (or transformation). Accumulation does not involve a change in the chemical and is simply a storage process. Degradation involves a breakdown or change in the chemical. Although, in general, the compound or compounds produced by transformation processes are less toxic, this is not necessarily true in all cases. Therefore, transformation does not necessarily imply the detoxification of a hazardous material.

4.0 GENERAL REFERENCES

Environmental Protection Agency (EPA), Water Related Environmental Fate of 129 Priority Pollutants. Washington, D.C., EPA 440/4-79-029 December (1979)

Environmental Protection Agency (EPA), Superfund Exposure Assessment Manual. Office of Remedial Response, USEPA Washington, D.C., EPA 540/1-88-001, April (1988)

Lyman, W.J., Reehl, W.F., and Rosenblatt, D.H., *Handbook of Chemical Property Estimation Methods: Environmental Behavior of Organic Compounds.* McGraw-Hill Book Co., New York. (1982)

Suffet, I.H., *Fate of Pollutants in the Air and Water Environments*, Wiley-Interscience, John Wiley and Sons, New York (1977)

Thibodeaux, L.J., *Chemodynamics: Environmental Movement of Chemicals in Air, Water, and Soil,* Wiley-Interscience, John Wiley and Sons, New York (1979)

Tinsley, I.J., *Chemical Concepts in Pollutant Behavior,* Wiley-Interscience, John Wiley and Sons, New York (1979)

3.

REGULATORY TOXICOLOGY

Toxicology is the study of the adverse effects (toxicity) of chemicals. As noted in Chapter 1, toxicology is concerned with answering the questions "What adverse effects are associated with exposure to a particular chemical?" and "What doses cause these effects?" Regulatory toxicology is that branch of toxicology concerned with ensuring that exposure to particular chemicals does not adversely affect people or the environment. Regulatory toxicology addresses the need to protect people from adverse effects in the face of uncertainty about the actual dose that is likely to affect any particular individual.

Toxicological data error is probably one of the largest sources of uncertainty in health risk assessments. As the EPA noted in its guidelines for carcinogenic risk assessment (13):

> There are major uncertainties in extrapolating both from animals to humans and from high to low doses. There are important species differences in target site susceptibility. Human populations are variable with respect to genetic constitution, diet, occupational and home environment, activity patterns, and other cultural factors.

Despite this uncertainty, toxicity criteria are often presented as precise, accurate values. This illusion of accuracy is exacerbated by the fact that the criteria are generally developed by a regulatory agency's expert panel (e.g., the EPA Reference Dose (RfD) Working Group for non-carcinogenic effects). However, the values produced by these toxicologists represent only a consensus on the dose unlikely to cause an effect. In many cases there are reasonable arguments that doses that are orders of magnitude lower would be safe or that the chemical may not cause the effect in humans at all. The primary purpose of this chapter is to describe the types of toxicity criteria used in hazardous waste risk assessment and to provide an understanding of the uncertainty inherent in these values. A secondary function of the chapter is to briefly outline the approaches used to collect and assess toxicology data and to develop toxicity criteria.

It is particularly important to understand the sources and magnitude of the uncertainty incorporated into the toxicity criteria because the false impression of accuracy can lead to illogical decisions in site remediation. For example, two groundwater samples, one with a concentration of some chemical slightly above an established criterion and the other slightly below may be treated differently when in fact the uncertainty in the toxicity-based criterion is over a factor of 100. As another example, considerable debate occurred at the EPA over whether 0.026 $(mg/kg/day)^{-1}$ or 0.029 $(mg/kg/day)^{-1}$ was the more appropriate cancer slope factor for benzene, when in fact, the difference between the two was scientifically indiscernible (20).

1.0 INTRODUCTION

In order to evaluate the potential for a chemical present at a hazardous waste site to adversely affect human health, information is needed on both the type of effects caused by the chemical and the dose required to cause these effects. In many cases, the toxic effects associated with long-term exposure to low levels of a chemical differ from the effects associated with short-term exposure to high levels. The effects of low-level long-term exposure to chemicals are generally emphasized in risk assessment. Chemical effects associated with long-term exposure are separated into two broad categories -- carcinogenic effects and noncarcinogenic or systemic effects. This distinction is necessary because of assumed differences in the mechanisms of action associated with the two categories. For noncarcinogens, most authorities consider there to be a dose below which no effect will occur (i.e., a threshold level). For carcinogens, however, many scientists assume that any exposure poses some risk (albeit very small) and consequently, it is assumed that no threshold exists. There has been considerable debate on this issue and there are good arguments supporting a threshold for certain chemicals. However, the EPA continues to promote the no-threshold argument for carcinogens, in part based on the arguments put forth by Schneiderman and others in the mid-to-late 1970s, that variability in human response and in exposure patterns make it impossible to distinguish between threshold and non-threshold effects (21).

1.1 Toxic Effects

The adverse health effects of most concern in humans are those that cause death or that are irreversible and seriously impair the normal functioning of the individual. Cancer, because it is potentially fatal and because of the widely held belief that at least for certain types of carcinogens there is no safe dose, is usually the toxic effect of most concern. Mutagenicity, or genetic toxicity, is important because it is felt that an alteration in the genes of a cell may be the first step in cancer formation and may cause teratogenicity (birth defects). Reproductive toxicity (effects on the reproductive system) decreases the ability to produce viable young.

Chronic toxicity involves effects seen following long-term, often lifetime exposure to a particular chemical. Acute toxicity is the effect of very short-term, usually single dose, exposure to a material. Subchronic exposure involves the administration of a toxic agent for a time period between chronic and acute exposure. For chronic or subchronic toxicity, the effects of concern are serious irreversible effects such as organ or tissue dysfunction (e.g., the central nervous system damage observed in the mad hatter caused by chronic mercury exposure). Lethality is the acute toxic effect of primary concern as most other effects due to acute exposure are reversible.

1.2 Dose-Response Evaluation

The process of establishing the dose of a chemical that is expected to produce an adverse effect in an exposed human is a complex task that is based on both scientific and policy considerations. "Safe" doses were initially based almost entirely on reported results in exposed humans. Knowledge on the toxicity of a chemical was initially noted by scientists based on trials on themselves (and probably on their assistants) and human response was used to establish "safe" levels. The obvious downside of this type of scientifically valid approach is that it required humans to be exposed to toxic doses of chemicals.

As noted by Zapp (22), the first animals used in toxicity tests were probably used as poison testers in the courts of Europe. The use of animals as test systems for observing the toxic effects of chemicals has evolved considerably since that time. A key driving factor in increasing the need for and the use of animals was increased concern over effects that occurred only after a long period of exposure. In particular, the assumption by the EPA that no threshold existed for cancer led to the need for studies employing large numbers of animals in order to obtain sufficient statistical power to determine if the very low doses that were of concern posed a risk.

The need to make a determination regarding the potential for effects to occur outside of the observable range was one of the factors responsible for the development of the comingling of science and policy that is regulatory toxicology (23). The other key factor was the realization that doses safe for one person or animal species were not necessarily safe for another. In order to ensure that human health was adequately protected, regulators began working towards establishing "safe" levels of exposure to chemicals. Developing these "safe" levels required incorporation of safety factors to ensure that most people who might be exposed would be protected.

1.3 Toxicity Values for Hazardous Waste Sites

Toxicity values used for assessing risks at hazardous waste sites are generally established by the EPA or by another regulatory agency. EPA (3) has established a hierarchy of approved sources of toxicity information for EPA site work, and these sources are often used by state agencies as well. Toxicity values that have been validated by an intraagency EPA toxicology workgroup are available on the Integrated Risk Information System (the IRIS database) (12) and are to be used preferentially to all other values at EPA-lead sites. If toxicity information for a particular chemical is not presented in IRIS, values presented in the EPA Health Effects Assessment Summary Tables (HEAST) (24) should be used. Finally, if toxicity values are not available from these sources, values from other EPA sources or from other agencies may be used as toxicity values. Some states do have toxicity values that differ from the EPA and in these states it may be necessary to use the state-approved values. In all cases, the source of the toxicity values used for a particular chemical should be clearly stated.

In addition to toxicity values that specify a critical dose to an organism, regulatory agencies also have developed media-specific criteria. These values specify the allowable concentration of a chemical in a specific medium (air, water, soil) and are developed based on assumptions about the likely intake of the chemical from the medium of concern. In establishing media-specific toxicity criteria, an allowable exposure dose or target dose must be determined. For carcinogens, the target dose is generally a dose associated with a cancer risk range of one in ten thousand (10^{-4}) to one in one million (10^{-6}), with the 10^{-6} dose level generally used as the target dose (more on this in Chapter 5). For noncarcinogens, the verified reference dose (RfD) is used as the target dose.

2.0 PHARMACOKINETICS

The behavior of a chemical in the body, termed pharmacokinetics (literally, drug movement) has a critical impact on the potential for the drug to cause harm (25). Consequently, the study of a chemical's pharmacokinetics and in particular the differences in chemical behavior in different species and individuals is an important subdiscipline of toxicology. Pharmacokinetics is generally subdivided into absorption (how the chemical enters the body), metabolism (changes that occur to the chemical), distribution (where the chemical goes and where it is stored in the body), and elimination (how the chemical leaves the organism).

2.1 Absorption

In order for a chemical to be toxic, it must come in contact with tissue that it can affect. Some chemicals are site-of-contact toxicants that affect epithelial tissues such as the skin (e.g., the polycyclic aromatic hydrocarbons or PAHs cause skin cancer following dermal exposure). Other chemicals also act at the site of contact but internally, either in the lung or in the gastrointestinal (G.I.) tract. However, the vast majority of chemicals exert toxic effects only after absorption into the body, either through the skin, via the G.I. tract, or by inhalation and absorption through the lung.

Inhalation. The upper respiratory tract is the site of absorption of many airborne chemicals, especially those that are water soluble or highly reactive (26). The rate of absorption of chemicals in the lung is highly dependent upon the particle size and the solubility of the chemical at the surface of the cell membrane. Chemicals that damage the lung by irritation or actual necrosis may increase their absorption or the absorption of associated chemicals by either increasing the size range of particles that can be absorbed by the lung or by changing the solubility of the chemicals.

Dermal Absorption. The skin is composed of several layers of cornified cells that cover the germinal cells and the dermis (27). The absorption of chemicals through the skin occurs relatively rapidly for lipophilic (fat soluble) substances, but slowly for hydrophilic (water soluble) chemicals. If the skin is damaged, its permeability to toxicants increases greatly. Organic solvents such as gasoline and hydrocarbons remove the lipids from the skin, making it readily permeable. Agents that cause dermal irritation increase the likelihood of chemical absorption, as the permeability of the capillary bed of the dermis and the local epidermal blood flow increase. Agents that cause frank necrosis of the epidermis, such as strong acids and bases, drastically alter the skin's permeability as well.

Ingestion. Chemicals that are ingested may be absorbed throughout the alimentary tract. However, some portions of the alimentary canal are specialized for absorption, namely, the stomach and intestines. The rate of absorption is determined by the chemical nature of the compound, whether it is a particulate or in solution, and whether is it ionized or nonionized. The medium in which the chemical is ingested (e.g., water, corn oil, food, soil) can significantly affect absorption. Chemicals that change the chemical nature of other compounds or that affect the gastrointestinal lining may increase or decrease the absorption of toxic chemicals by the gastrointestinal tract.

2.2 Metabolism

Chemicals absorbed into the body may pass through the body unchanged, may be routed to various parts of the body and stored, or may be transformed (metabolized) in the body to either more or less active chemicals (metabolites). Some chemicals require enzymatic activation in order to bind to RNA, DNA, or proteins. Biotransformations of chemicals occur in the liver, kidney, lungs, intestines, skin, and other organs, with the liver the most important site of metabolic activity. The most common routes of metabolism involve oxidation, reduction, hydrolysis, and conjugation. The various reactions often occur simultaneously through the action of the microsomal mixed-function oxidases or sequentially through the action of microsomal and cytoplasmic enzymes that catalyze the conjugation of glucuronic acid, sulfate, glutathione acetate, and other moieties.

The microsomal mixed-function oxidase system, which normally metabolizes steroids and foreign compounds absorbed from the digestive tract, is one of the most important metabolic pathways for organic chemicals (25). The enzyme activity is mediated through the terminal oxidase, cytochrome P-450 and is activated by the presence of chemicals undergoing metabolism. Consequently, many compounds markedly increase the activity of microsomal mixed function oxidases and can increase their own or other chemicals' metabolism. There are at least two categories of inducers of mixed-function oxidase. The first, exemplified by phenobarbital, is composed of drugs and insecticides that increase the rate of metabolism of a wide variety of xenobiotics. The second group, exemplified by 3-methylcholanthrene, is composed primarily of polycyclic hydrocarbons that stimulate the metabolism of a more limited number of substrates. The number of chemicals known to induce the activity of microsomal mixed-function oxidase is extensive.

In general, the process of metabolism is designed to increase the rate of elimination of a chemical in order to facilitate its removal from the body. Consequently, in many cases, metabolism represents a detoxification mechanism. As an example, inorganic arsenic is fairly quickly methylated (a methyl group is added to the metalloid) and once methylated, is fairly rapidly excreted in the urine. Arsenic taken into the body in a methylated form (the arsenic in fish for example) is rapidly excreted and is not considered to be toxic. However, metabolism for many organic compounds can yield more toxic metabolites or can produce short-lived, toxic products (e.g., free radicals) in the process of transforming to a more stable and mobile compound. Consequently, metabolism can be a detoxification mechanism or can act to increase a chemical's toxicity.

2.3 Distribution

Once in the body, chemicals enter the blood stream and are generally carried throughout the body (26). Different chemicals will have particular affinities for specific locations in the body and will tend to concentrate in these tissues. For example, arsenic is distributed preferentially to the skin and lipophilic compounds tend to concentrate in adipose tissue and in organs that contain high levels of fat such as the brain. The distribution of a chemical in the body is usually at least somewhat predictive of the tissues that will be adversely affected by the compound because these tissues will have the highest concentrations of the chemical. However, in some cases (e.g., fat in adipose tissue) the chemical is simply being stored. The binding or storage of compounds in nontarget tissues such as adipose tissue may effectively sequester considerable amounts of chemicals and their unconjugated metabolites. However, tissue binding is generally reversible, and chemicals may be mobilized under appropriate conditions such as fasting or dieting.

2.4 Excretion

Clearance or removal of toxicants may occur by direct exhalation, excretion from sweat glands, elimination in fecal matter, and excretion via the kidneys. The major pathway for removal of most chemicals is via the urine but both fecal elimination and exhalation can be important for certain types of chemicals.

Renal excretion of chemicals that are absorbed into the blood stream and tissues is subject to the complex physiology of the kidney. For example, various drugs and metabolic conjugates are secreted by active transport across the tubular epithelium (NRC 1980). If polar metabolites of foreign chemicals are actively secreted by the kidney, inhibitors of mixed-function oxidase or conjugation reactions that produce these polar metabolites would prolong the half-life of the compound. Also, substances that are excreted by the same active transport mechanism compete with each other. The size of the dose, with possible saturation of renal transport, the extent of binding at sites on plasma proteins, and factors that limit the availability of essential cofactors (glycine, sulfate, and glutathione) limit the rate of renal elimination. Renal toxicants, such as lead, cadmium and mercury, cause severe damage that limits renal function. Factors that limit blood flow to the tissues may also limit the rate of elimination of toxicants bound in tissues.

Organic chemicals and metabolites formed in the liver may be secreted in the bile by active biliary transport mechanisms. Biliary excretion of high molecular weight compounds is enhanced by glucuronidation (25). Particularly for compounds that undergo metabolism by glucuronidation, elimination in fecal matter can be an important pathway.

Elimination of chemicals from the body via the lungs is an important route for some volatile compounds and for certain metabolites of these compounds, specifically CO_2. Agents that increase the rate of respiration, irritate the pulmonary epithelium, or cause bronchiolar constriction or even pulmonary fibrosis would influence the elimination of gaseous toxicants (26).

3.0 DEVELOPMENT OF TOXICITY CRITERIA

As a chemical moves through the body, it may exert a harmful effect on tissues or organs that it or its metabolites contact. Toxicology is focussed on determining how the chemical causes this effect and how much of the chemical is required to cause the effect. In order to make these determinations, exposed human populations and animals-administered doses of the chemicals are studied. The results of these studies are collected, compared, evaluated, and used as the basis for decisions about the toxicology of the chemical.

3.1 Test Systems

The biological response to chemical exposure in humans is studied using three general methods. Epidemiologists study the response of human individuals or populations exposed to the chemical of concern and compare effects on this group to effects on a similar but unexposed group. Animal toxicologists expose laboratory animals to chemicals, observe the chemical related effects, and attempt to extrapolate the effects observed in animals to potential effects on man. Increasing emphasis is being placed on a third area of toxicology, the use of animal, plant, and microbial systems in short-term tests to evaluate specific effects of chemical on these systems.

There are advantages and disadvantages to all methods. Studies of exposed humans are usually considered qualitatively most dependable as there is no need to make the assumption that humans will be affected in the same manner as animals. However, there are obvious ethical problems with knowingly exposing humans to toxic doses of chemicals to evaluate their effects. Studies of humans previously exposed to chemicals usually lack reliable quantitative information on chemical exposure and have potential for bias and confounding. Also, it may be difficult to get a large enough sample size to observe a significant increase in a particular toxic effect even if, in fact, the chemical does cause the effect.

Animal studies, on the other hand, are more easily controlled and therefore are good sources of quantitative data. However, exposure to toxic materials often affects different species in a dissimilar fashion, both qualitatively and quantitatively (28,29). Consequently, simply because a toxic response is observed or not observed in an animal species is not positive evidence that the same effect will occur or not occur in humans. Therefore, good quantitative dose-response information in one species may be totally inapplicable to a different species. To attempt to overcome some of these potential problems, toxicologists generally try to test more than one species of animal for a particular response in order to ascertain that the effect is not species specific. Pharmacology of the chemical in different species can also be reviewed to look for differences. For example, arsenic binds tightly to the red blood cells (RBCs) of rats but apparently not to the RBCs of man or of most other species (30).

The testing of chemical effects in animals has evolved to a fairly sophisticated process, requiring very careful control on animal living conditions, diet, and care, and generally conforming to established regulatory requirements for number of animals and exposure dose levels (31,32,33). The types of animal studies conducted generally fall into categories of acute toxicity testing (usually single doses are administered), short-term testing (further subdivided into subacute studies, generally lasting around 14 days and subchronic assays lasting 30 to 90 days) and chronic bioassays that generally last for close to the full lifetime of the animal.

Short-term tests also easily controlled, are generally relatively inexpensive, and in many cases can provide good information on the potential for an effect to occur. However, it is even more difficult to extrapolate from bacteria to humans. Consequently, the short-term test systems are primarily used to provide supporting evidence in criteria development. A fourth type of toxicity testing, the study of structure-activity relationships (i.e., attempting to predict how one chemical will behave based on the behavior of another chemical with similar features) is an area of toxicology that is currently primarily used to provide supporting information for criteria. Structure activity information and short-term tests are currently receiving most use as screening tools in new chemical development to evaluate toxicity prior to initiating much more costly animal bioassays. However, as the ability to accurately predict outcomes based on the tests increases, and as the knowledge base for these tests increases, their role in developing regulatory criteria is likely to expand.

3.2 Procedures Used To Standardize Dose

The most common source of toxicity information used in developing criteria for risk assessments is animal bioassays. These bioassays are conducted on a wide range of animals using various dosing schedules, administration procedures, and exposure duration. However, the final toxicity criteria need to be expressed in units of mg chemical/kg human bodyweight/day. Consequently, procedures are required to convert the doses in animals to an equivalent dose in humans. These conversions, of course, require assumptions about the equivalence of species and as such introduce substantial uncertainty into the risk assessment.

Laboratory tests of the toxicity of chemicals are often conducted using different routes of administration. The test compound can be administered orally in the food, drinking water, or by gavage; it can be given by injection; or the compound can be incorporated into the air the animal breathes and thereby administered by inhalation. The following procedures can be used to standardize units of dose. As noted above, toxicity criteria are expressed in units of mg chemical/kg body weight/day or simply mg/kg/day.

Feeding Studies. For those studies in which the dose is expressed as part-per-million (ppm) or percentage of the diet, the dose is first converted to units of milligrams of compound per gram of food:

Eq (1) $$X \text{ ppm} \times 1/1000 = y \text{ mg/g}, \text{ or}$$
$$X\% \times 10 = y \text{ (mg/g)}.$$

Then, the following conversion to administered dose is made:

Eq (2) $$\frac{\text{mg compound}}{\text{g food}} \times \frac{\text{g food consumed}}{\text{day}} \times \frac{1}{\text{body weight of animal (kg)}} = \text{dose(mg/kg/day)}$$

The average body weights and grams of food consumed per day for various species are shown in Table 1. These values are approximate. Actual data are commonly available in the study report and if available, should be used.

Drinking Water. For those studies in which the dose is administered in drinking water, the dose is expressed in mg/liter of drinking water (note: 1 part per million or ppm = 1 mg/liter) and the following conversion is made:

Eq (3) $$\frac{\text{mg compound}}{\text{liter drinking water}} \times \frac{\text{liters water consumed}}{\text{day}} \times \frac{1}{\text{body weight (kg)}} = \text{dose (mg/kg/day)}.$$

TABLE 1

Food Consumption in Various Species

Species	Body Weight (kg)		Food Consumed Per Day (g)	
Source:	(34)	(35)	(34)	(35)
Mouse	0.02	0.035	3.	5.
Chicken	0.40	0.12	50.	10.
Rat	0.40	0.45	20.	10.
Guinea pig	0.75	0.43	30.	12.
Rabbit	2.0	3.7	60.	180.
Dog	10.0	14.0	250.	400.
Monkey	5.0	3.5	250.	150.
Man	70.0	--	1500.	--

TABLE 2

Minute Volumes for Selected Species

Species	Body Weight (kg)	Volume of Air Breathed in 1 Minute (liter)
Mouse	0.02	0.023
Rat	0.40	0.0735
Hamster	0.50	0.054
Guinea pig	0.75	0.16
Dog	10.0	5.2
Monkey	5.0	0.86
Man	70.0	10.4

Source: (34)

Volumes of drinking water consumed for various species are estimated to be: mouse = 0.005 liters, rat = 0.100 liters, and human = 2.0 liters (34). Body weights are those shown in Table 1.

Inhalation. For those studies in which the concentration of the compound is expressed in ppm or mg/m3 in air (note 1 ppm_{air} is not equal to 1 mg/m^3), all doses are first converted to mg/liter in air:

Eq (4) $$\frac{\text{ppm x molecular weight of the compound}}{24{,}450} = \text{mg/liter}$$

and

$$\frac{mg/m^3}{1{,}000} = \text{mg/liter.}$$

Once inhaled concentrations are expressed in mg/liter of air, they are converted to mg/kg/day as follows:

Eq (5) $$\text{Air concentration (mg/liter) x } \frac{\text{liters of air breathed}}{\text{minute}}$$

$$\text{x } \frac{\text{body weight}}{\text{(kg)}} \text{ x } \frac{\text{minutes exposed}}{\text{day}} = \text{dose (mg/kg/day)}$$

Minute volumes for various species are given in Table 2.

3.3 Systemic Effects

Once the dose is expressed in units of mg/kg/day, the next step is to develop the criteria by adjusting the study dose to ensure humans are protected (12,36,37,38,39). For noncarcinogens, the primary approach used to develop health criteria involves determining the highest dose that is not expected to result in an adverse effect (the No Observed Adverse Effect Level or NOAEL) from toxicity studies in humans or animals. Typically, the study giving the lowest NOAEL is used (or in toxicological jargon, the most sensitive sex of the most sensitive species is used). The

No Observed Effect Level or NOEL was sometimes used in the past but has generally been replaced by use of the NOAEL. If a NOAEL is unavailable from the most sensitive study, then the lowest dose tested that resulted in an adverse effect (the Lowest Observed Adverse Effect Level or LOAEL) is used. The NOAEL or LOAEL is then divided by protection factors (currently called uncertainty factors by the EPA and formerly referred to as safety factors) to obtain a reference dose (RfD; formerly called an allowable daily intake or ADI) in units of mg (chemical)/kg (bodyweight)/day (40). Protection factors of 10 are commonly used, as necessary, to account for potential differences between the animal species used in the toxicity study and humans in sensitivity to the compound, for differences in duration of exposure, and for potential differences among people in their response to the chemical (41). A protection factor of between 1 and 10 is also used if a LOAEL rather than a NOAEL serves as the basis of the RfD to adjust for the severity of the observed effect. A modifying factor can also be added to account for other concerns.

As an example of the development of an RfD, carbon tetrachloride administered to 1 kg rats at a dose of 1 mg/day for 5 days/week for 12 weeks showed no effects while in the same study higher doses produced liver lesions (24). The daily dose of 1 mg/kg x 5/7 = .71 mg/kg/day is the NOAEL. This NOAEL is divided by a factor of 10 because the study was for less than a lifetime, another factor of 10 to account for interspecific differences in response to the chemical, and a third factor of 10 to account for variability among humans in response to chemicals. Therefore, the final RfD of 7 x 10^{-4} mg/kg/day for carbon tetrachloride incorporates a protection factor of 1000.

The EPA notes that the RfD is an estimate of the daily chemical exposure dose that is considered unlikely to cause adverse effects during a lifetime of exposure (12). The RfD is a benchmark value, and while doses below the RfD are unlikely to be of public health concern, it is possible that a hypersensitive individual could be adversely affected at this level. Similarly, exposure to doses even significantly above the RfD indicates that an inadequate margin of exposure (MOE; formerly referred to as the margin of safety) exists, but should not be construed to mean that an adverse effect is imminent or that the exposure is by definition "unacceptable" or "of concern." The MOE can be defined as the ratio of the no observed adverse effect level or lowest effect level (if no NOAEL is available and the observed effect was minor) to the estimated exposure dose. The MOE gives an indication of the margin of safety between estimated exposure and actual levels of concern.

It should be noted that EPA commonly refers to the RfD as having perhaps an order of magnitude of associated uncertainty (12). In reality, these values tend to have a much higher level of uncertainty. Uncertainty factors of between 10 and 1,000 are commonly added to the values to ensure that the values are health protective. Even ignoring the question of whether the effect observed in one species will even occur in another, uncertainty in the RfDs is clearly greater than an order of magnitude.

3.4 Carcinogenicity

For carcinogens, results from either human epidemiological studies or chronic animal carcinogenicity bioassays are used in a mathematical model to estimate the slope of the line relating exposure (dose) to effect (response) (42). The mathematical model that is most often used by EPA in developing cancer slope factors is the linearized multistage model developed by Armitage and Doll and adapted for computer use by Crump and colleagues as GLOBAL 82 (43). Recently, models that purport to account for pharmacokinetic information are being used more frequently (i.e., the so called physiologically based, pharmacokinetic or PB-PK models; 44,45,46). However, the underlying uncertainty in any of the cancer risk assessment models is large and should not be ignored. The science and art of quantitative risk assessment as it applies to developing cancer slope factors is well beyond the scope of this book. Suffice it to say that the values established by the various researchers (and often argued about extensively) are usually at best only rough order-of-magnitude estimates of possible risk.

The numerical value that is derived from the cancer risk models is called the cancer slope factor (formerly the cancer potency factor) and is expressed in units of $(mg/kg/day)^{-1}$. The cancer slope value can be multiplied by an individual's lifetime average dose of a compound in mg/kg/day to estimate the probability of developing cancer associated with that exposure. The cancer slope factors are generally developed using conservative assumptions and the upper 95% confidence limit (UCL) of the maximum likelihood estimate (MLE) of the line relating dose to response. For these reasons the EPA (12) notes that the slope factors are rough but plausible estimates of the upper limits on lifetime risk. Actual risks estimated using these values are unlikely to be higher but could be considerably lower.

In addition to presenting cancer slope factors, the EPA currently lists unit risks for inhalation and drinking water exposure to carcinogens. Unit risks are media-specific values that express the potency of a carcinogen in terms of risk per unit of the particular medium (e.g., risk per ug/liter for drinking water and risk per ug/m^3 for inhalation exposure). Implicit in the unit risk numbers is the assumption that the exposed individual weighs an average of 70 kg and drinks 2 liters of water per day or inhales 20 m^3/day of air. Risks are determined by multiplying the unit risk value by the lifetime average concentration of the carcinogen in the particular environmental medium.

3.5 Reproductive/Developmental Effects

Toxicants that affect the reproductive system or that cause adverse effects on the developing fetus can generally be evaluated using the same approaches outlined above for other systemic effects (3,12). In general, a dose from a study of the chemical's effects is divided by various protection factors to arrive at a reference dose. However, for developmental toxicants (chemicals that affect the developing fetus), the critical exposure period is typically pregnancy and consequently, exposure which occurs during this time period, (and not necessarily a full lifetime) is of concern. RfDs for developmental toxicity are generally developed taking this shorter time period into account.

4.0 ISSUES IN TOXICOLOGY

Standard risk calculations do not address all the issues that may be of concern at a site. Short-term release of chemicals may be a concern to local residents but generally poses less risk than longer-term exposures so it is not discussed in detail in many risk assessments. Many diseases, including cancer and reproductive effects are not rare events and an understanding of background incidence rates is important. Finally, chemicals generally occur as mixtures in the environment and the potential impact of multiple chemical exposures needs to be considered.

4.1 Short-term Effects

Risk assessment at hazardous waste sites commonly focuses on the effects of extended or chronic exposure to chemicals as does most current risk assessment for hazardous materials. However, residents living in the vicinity of waste sites, industrial plants, and other potential sources of pollution often express concern about other toxic effects of the chemicals to which they are being exposed. In particular, these residents often complain about adverse effects associated with short-term exposure to airborne materials such as eye- or mucous-membrane irritation.

Ideally, toxicity values from short-term studies are used to assess the risks associated with a short-term elevation in the levels of airborne chemicals. Such short-term exposure may be associated with an inversion layer of short duration or with an accidental release of a compound. However, few regulatory agencies have established health criteria for short-term exposure to chemicals. The EPA Office of Drinking Water has developed one (1) day and ten (10) day Health Advisories for a number of chemicals in water (47). NIOSH has established levels of chemicals that are immediately dangerous to life and health (IDLH) for acute occupational exposure. Neither of these criteria are appropriate for assessing potential health risks to human populations exposed to airborne chemicals. The Threshold Limit Values (TLVs) developed by the American Conference of Governmental Industrial Hygienists (ACGIH) to protect workers exposed for 8-hour workdays to chemicals in the workplace can be used to derive health criteria that will be protective for populations that may be exposed for short periods of time (several hours to several weeks) to chemicals released into the environment (48,49). In order to account for differences in exposure duration and susceptibility between the worker population and the general population, the TLVs are divided by a factor of 280. This includes factors to adjust for differences in exposure period (the average worker is assumed to breathe 10 m^3 air during an 8-hour workday for 5 days per week whereas the average individual breathes 20 m^3 air during a 24-hour day for a 7 day week), and individual susceptibility (a factor of 10 is used to account for the fact that many of the ACGIH TLVs are minimal effect levels rather than no effect levels). Sensitive individuals in an environmentally exposed population may be affected by chemicals at or below the TLVs. Consequently, it is imperative that the basis for the TLV be reviewed prior to use to ensure that an appropriate protection factor has been added to protect potentially sensitive individuals.

Another approach that can be used to evaluate the potential short-term effects of chemicals in air is to use the RfDs derived for long-term exposure. If exposures do not exceed these criteria, no further action is necessary. If the RfDs are exceeded, then further evaluation of data on the subset of chemicals causing concern may be required.

4.2 Carcinogenicity

As previously noted, cancer risks are assessed differently than noncancer risks based on the assumption that no threshold exists for cancer (50). Consequently, a first step in the toxicological evaluation process is a determination as to whether or not a chemical is a carcinogen. Information from animal bioassays, short-term tests, and information on related compounds is evaluated to make the determination. A classification system has been developed by the EPA (12) for reporting the weight of evidence for carcinogenicity for chemicals. This classification scheme is based on previous approaches developed by Griesemer and Cueto (51) and the International Agency for Research on Cancer (IARC) (52). In the EPA categorization approach, Group A carcinogens include chemicals that are considered to be carcinogenic based on sufficient evidence of carcinogenicity from epidemiological studies. Chemicals in Group B1 - Probable Human Carcinogens, are considered to be carcinogens based on limited evidence from human studies and sufficient evidence of carcinogenicity from animal studies. Group B2 - Probable Human Carcinogens includes chemicals for which there is sufficient evidence of carcinogenicity in animals but no evidence or insufficient evidence from human epidemiological studies. Chemicals are classified in Group C - Possible Human Carcinogens based on limited evidence of carcinogenicity from animal studies. Group D - Not Classifiable as to Human Carcinogenicity includes chemicals that have inadequate data on carcinogenicity. Finally, chemicals that have shown no evidence of carcinogenicity in adequate human or animal studies are classified in Group E - Evidence of Noncarcinogenicity in Humans.

The nonthreshold theory of carcinogenesis is based on the belief that cancer is a multistep process, with the first step, termed initiation, commonly involving damage to DNA (43). It is assumed that initiation can occur at very low doses and once it has occurred, the cancer can continue to develop even in the absence of the initiating agent. It should be noted that not all DNA damage is irreversible and not all initiated cells will necessarily move into the next stages. In addition, carcinogens that act at later points in the cancer process may only be effective at high doses (i.e., there may be a threshold for these compounds) (53). However, EPA and other U.S regulatory agencies such as FDA currently assume that no threshold dose can be identified for most carcinogens.

A final note on cancer risks is that cancer is a rather common disease in the normal population. Approximately 30% - 40% of the normal U.S. population will develop cancer in their lifetime and about half of these people will die from the disease (54). This rather large incidence means that at essentially any hazardous waste site, some nearby residents or former workers are likely to have cancer or to have died from cancer. Simply observing these effects in these people does not necessarily imply a link to the site. Of course, conversely, observing only a small number of cancers in a population does not mean that these cancers are not site related. Risk assessors need to be aware that proving or disproving a link between observed cancers and a site is likely to be impossible. In most cases, all that can be done is to provide available data on cancer incidences in the general population in a manner that is sensitive to the concerns of local residents.

4.3 Reproductive/Developmental Effects

The effects of chemicals on the ability to produce viable young is generally not a major issue in evaluating the risk at hazardous waste sites. In general, if chemical concentrations are high enough to pose reproductive/developmental risks, the concentrations are also high enough to be of concern due to systemic or carcinogenic effects. Having noted that these effects are seldom driving forces in the risk assessment, it is important to note that they are often a major concern to local area residents. Consequently, at least some discussion of the issues should be provided in risk assessments of sites at which chemicals with reproductive or developmental effects are detected.

Reproductive effects, including effects on male and female reproductive systems and fetal toxicity, can occur as a result of exposure to chemicals that act specifically on the reproductive system or as a result of exposure to chemicals that severely stress the adult. 1,2-Dibromo-3-

chloropropane (DBCP), for example, adversely affects male reproduction (causes sterility) at doses that do not produce other signs of serious toxicity (55). Ethylbenzene on the other hand, is an example of a compound that does not produce adverse effects on the developing young until effects are seen in the parent, generally at high levels that are unlikely to be encountered at a hazardous waste site.

Teratogenicity is the ability of a chemical to produce birth defects in animals exposed in utero (56). Concern over teratogenic effects has been stirred by the thalidomide and diethylstilbestrol (DES) incidences in children of women administered these agents therapeutically. Thalidomide was used to limit vomiting and nausea during pregnancy and had no obvious toxic effects at therapeutic doses. However, thalidomide was shown to produce a rare form of very severe limb malformation in some children exposed in utero to the drug. DES produces a type of rare tumor of the vagina in women whose mothers had taken the drug during pregnancy to prevent threatened miscarriages.

The teratogenic effects associated with exposure to these chemicals occurred in only a portion of the population exposed. The detection of teratogenic effects in these cases was possible in large part because the effects produced were rare events and therefore notable to researchers. A comparable increase in a more common malformation may not have been detected. For a teratogenic effect to occur, exposure has to take place at the exact time when the target tissue (of the offspring) is under development. If the target organ is not undergoing development at the time the teratogen is administered, effects will not occur.

As noted above, reproductive and developmental effects of chemicals are generally not of primary focus at hazardous waste sites because other effects such as carcinogenicity are of concern at lower levels. The exception is chemicals whose principal toxic effect is reproductive impairment. For risk assessment purposes, reproductive impairment is generally treated like a systemic effect. Teratogenicity is generally not likely, as the chance is low that a pregnant woman will come in contact with a high enough dose of a site chemical at the exact point in time when that chemical could cause its effect. However, at sites near residential areas or where exposure could be high enough to produce teratogenic effects, teratogenicity should be considered.

It is not uncommon to learn of miscarriages and birth defects in women residing near hazardous waste sites. While these occurrences are certainly distressing, the incidence of miscarriages and birth defects in the general population is higher than commonly expected, and the simple fact of occurrence does not necessarily imply a link to the site. Incidences for miscarriages in the U.S. population is 15% of all recognized pregnancies; 7% of all babies are born prematurely, and 8% of all babies have major or minor birth defects that are evident within the first year (57). Considering that people near a hazardous waste site are more likely than the average population to be looking for such events, it is not unexpected that these normal but undesirable events would be noted and anecdotally linked with the site.

4.4 Chemical Mixtures

The waste stream emitted from a hazardous waste site is likely to contain more than just a single chemical. These chemicals may not affect each other or may react to increase or decrease their respective toxicity (58,59,60). Chemicals can interact at different sites and for different reasons. Some interactions will affect the bioavailability of the compound to the target site. Other interactions may involve chemical reactions that transform the compounds involved. There are four main types of chemical interactions: addition, synergism, potentiation, and antagonism (61). Chemicals that interact so that the combined effect of the agents is equal to the sum of the effect of each chemical given alone are said to act additively. Chemicals are said to act synergistically if their combined effect is greater than the sum of the effects of the individual chemicals. Potentiation refers to a case in which a chemical that has no effect on a particular organ or tissue enhances the effect of another chemical on that system. Chemicals that interact so that the combined effect is less than the sum of the individual effects are said to interact in an antagonistic manner. In complex mixtures all of these effects may be exhibited.

There are several types of antagonistic interaction. Chemical antagonism involves a chemical reaction between the interacting chemicals that produces a less toxic product. Dispositional antagonism involves an effect of the disposition of the chemical in the body such that the amount of a toxic agent reaching a target tissue is reduced. Two chemicals that have opposite effects on the same target organ are called functional antagonists. Receptor antagonists or blockers are chemicals that compete antagonistically for the same binding site or receptor.

Although it is possible that chemicals will all act independently of each other it is more likely that at least some of them will interact. As noted above, such interactions may increase, decrease, or have no effect on the toxicity of the individual chemicals involved. The EPA (61) recommends that in the absence of data to the contrary, chemicals be considered to interact additively but that the site and mechanism of action of individual chemicals should be considered in assessing potential risks associated with exposure to chemical mixtures.

5.0 EXAMPLE TOXICITY PROFILES OF COMMON SITE CHEMICALS

Toxicity profiles describing the toxicity of a chemical and key criteria are generally required as part of most risk assessments for hazardous waste sites. A profile has probably been prepared for almost every chemical detected at a waste site and if at all possible, attempts should be made to locate an existing profile rather than writing a new profile. However, it may be necessary to write a profile in some cases. Consequently, this section provides sample toxicity profiles for several chemicals that are commonly detected at hazardous waste sites. The profiles provided include two examples of short, rather straight-forward profiles (for benzene, a carcinogen and for trans-1,2-dichloroethylene [trans-1,2-DCE], a noncarcinogen). Most toxicity profiles in risk assessments can take this form and are used to provide the toxicity criteria for the chemical and to present a limited amount of information about the effects of the chemical. A more detailed profile is next provided for lead. The potential for lead toxicity is evaluated based on the body burden of lead as measured in the blood, and therefore the toxicity profile for lead develops target levels in environmental media (in this case soil) that would produce lead levels in the body that are below the target body burden. The final profile, for polychlorinated dibenzo-p-dioxins and furans (PCDDs/PCDFs) provides an approach to addressing compounds that occur in the environment as complex mixtures.

5.1 Example Profile - Benzene

Benzene is a recognized human carcinogen, causing leukemia in occupationally exposed workers (12,55). Benzene has also been associated with the induction of aplastic anemia and with other blood synthesis disorders. Exposure to high levels has been associated with cardiac sensitization and cardiac arrhythmia. The EPA (12) has determined that benzene is a known human carcinogen (Weight of Evidence Category A) and has calculated a cancer potency factor for benzene of 0.029 $(mg/kg/day)^{-1}$.

5.2 Example Profile - trans-1,2-Dichloroethylene

There are no reports of carcinogenic or teratogenic activity by trans-1,2-dichloroethylene (trans-1,2-DCE) in animals or humans (55). trans-1,2-DCE has anesthetic properties and has been found to cause nausea, vomiting, weakness, and cramps in humans. Compound-specific information concerning the toxicity of trans-1,2-DCE is not adequate, and health-based criteria are derived from data for 1,1-dichloroethylene. The EPA (12) has derived an oral RfD of 0.02 mg/kg/day but has not developed an inhalation RfD. The oral RfD will be used to estimate both ingestion and inhalation risks associated with human exposure to this compound. A maximum contaminant level goal (MCLG) and a maximum contaminant level (MCL; also known as the Primary Drinking Water Standard) of 0.1 mg/liter have been established (47).

5.3 Example Profile - Lead

Lead, at relatively low exposure levels, has been shown to cause adverse neurological effects in humans, especially young children and fetuses. Of particular concern are the effects on learning ability that have been associated with very low (e.g., 10-20 ug/dl) blood lead levels (62). Oral ingestion of some lead salts has been associated with increased renal tumor frequencies in rats. The EPA (12) has classified lead as a Probable Human Carcinogen (Group B2) based on these results. The EPA has also noted that the available data provide an insufficient basis on which to regulate carcinogenic lead salts and no quantitative estimate of the excess cancer risk has been provided.

Most health effects associated with exposure to lead are reported in terms of a given amount of lead in the body, with the most commonly used measure of body lead levels being the concentration of lead in the blood. For this reason, the risks associated specifically with lead are analyzed using a different approach than used for other systemic toxicants.

The EPA (12) has determined that low levels of blood lead, in the 10-15 ug/dl range, may be of public health concern. These levels of concern are only marginally above the national average levels of 9.2-10 ug/dl (63). Because background levels, particularly in urban environments, are already high (64), it appears that <u>any</u> increase in blood lead levels would be of concern. CDC (65) noted that levels of lead in soil of from 500-1000 mg/kg may lead to increases in blood lead levels in children. These soil levels are consistent with data from the EPA (66) that suggest that in urban areas, the slope of the line relating blood lead levels to soil lead levels ranges from 0.6 to 2.2 ug/dl per 1000 mg/kg. In other words, the EPA data suggest that a soil lead concentration of 500 mg/kg, approximately 500 mg/kg above the national background soil lead level of 16 mg/kg reported by Shacklette and Boergnen (67), would be associated with an increase in blood lead in children of from 0.3 to 1.1 ug/dl. Because of statistical variation and uncertainties in measurement, it would be difficult to definitively attribute an increase of less than 1 ug/dl to a specific source. Therefore, based on a desire to prevent any increase in blood lead, a cleanup criteria of 500 mg/kg would be advisable for residential areas. As noted in EPA (63), a large number of factors can influence the soil lead-to-blood lead relationship and the different conditions associated with the different studies would probably influence the derived blood lead-to-soil lead relationships.

Because adults are more tolerant to the adverse effects of lead, higher soil lead levels may be tolerated at an industrial site where only adult exposure is likely. Blood lead levels as low as 40 ug/dl have been associated with toxicity in occupationally exposed workers. The Occupational Safety and Health Administration (OSHA) currently uses a blood lead threshold level of 50 ug/dl as a maximum permissible exposure level and also uses a blood lead level of 30 ug/dl as a recommended action level. For hazardous waste site evaluation, the OSHA action level of 30 ug/dl, which is below levels shown to cause effects in adults, should be used as a target criteria.

Baseline blood lead levels and the relationship between soil lead levels and blood lead levels must be determined before allowable soil lead levels can be established. The EPA (63) reported that geometric mean blood lead levels in 1978 for adults were 10 ug/dl and 15 ug/dl for women and men, respectively. The EPA estimated that these mean levels will be reduced to a value of around 4-5 ug/dl by 1990 as a result of the phase-out of leaded gasoline and decreases in the amount of lead in the diet. This value seems likely to be overly optimistic for many urbanized areas and a blood lead level around 10 ug/dl is probably more realistic as a baseline value.

The EPA (63) reviewed numerous studies on the blood lead-to-soil lead relationship. Based on these studies, the EPA estimated that the slope relating blood lead to soil lead levels was 0.002 ug/dl per 1 mg/kg. The EPA also calculated the relationship and determined each 1 mg/kg increase in soil lead concentration was associated with a 0.005 ug/dl increase in blood lead. Using the calculated ratio, the expected baseline blood lead level of 10 ug/dl, and the OSHA action level of 30 ug/dl as a target blood lead criteria, an action level (AL) for soil lead can be calculated using the formula:

Eq (6) $AL = 1 \text{ mg/kg } (0.005 \text{ ug/dl})^{-1} \times (30 \text{ ug/dl} - 10 \text{ ug/dl})$

Based on these calculations, an AL for soil lead of 4,000 mg/kg would be allowable in areas where exposure could be limited to adults. If the empirically derived blood lead-to-soil lead ratio is used, a value of 10,000 mg/kg would be allowable.

The values of 4,000 mg/kg or 10,000 mg/kg are probably sufficiently protective for adults under most conditions. However, these values do not consider the potential that lead in certain industrial settings (such as near lead smelters or lead battery reclaimers) might be more bioavailable than lead in the studies used as the basis for determining the blood lead-to-soil lead relationships and that children may be exposed to lead at industrial sites if it is carried home on workers' clothes. Because of these considerations, a value of 1,000 mg/kg is recommended as a cleanup level for lead at industrial sites. However, hotspots containing somewhat higher levels are unlikely to pose a health risk as long as the overall site and in particular, high use areas, contain levels averaging below 1,000 mg/kg.

5.4 PCDDs/PCDFS

Polychlorinated dibenzo-p-dioxins (PCDDs) and polychlorinated dibenzofurans (PCDFs) are products of the incomplete combustion of chlorinated materials and are also formed from polychlorinated biphenyls (PCBs) and pentachlorophenol (PCP) at high temperatures. Several studies have indicated that these compounds are ubiquitous in the environmental (albeit at very low levels). For example, Travis and Hattemer-Frey (68) summarized data from several other researchers and noted that PCDD/PCDF were present at levels of over 1 ug/kg in suburban U.S. areas with no known local source. [Note: While not strictly part of the toxicity profile, information on expected background levels of the compound can be quite useful in giving perspective on concentrations detected at a site.]

PCDDs/PCDFs are believed to share a common mechanism of action, with individual congeners differing widely in potency. The most biologically active congeners tend to be chlorinated at the 2,3,7,8-position, and 2,3,7,8-tetrachlorodibenzo-p-dioxin is commonly considered to be the most potent of the PCDD/PCDF compounds. For regulatory purposes, the relative carcinogenic potencies of the other PCDDs/PCDFs are generally estimated based on a comparison of their overall toxicity with that of 2,3,7,8-TCDD (69).

TCDD has been shown to be carcinogenic when administered orally or dermally to mice or rats (Kociba et al. 1978; NTP 1982a,b). The EPA Carcinogen Assessment Group calculated a cancer potency factor, which is the upper 95th percentile confidence limit of the slope of the line relating response to dose (or risk to exposure), of 1.56×10^5 $(mg/kg/day)^{-1}$ for TCDD (12). The EPA is reevaluating this cancer slope factor based on a consideration of the likely mechanism of action of TCDD.

TCDD Toxicity Equivalence Determination. Although a considerable amount of data is available on TCDD, less information is available on other PCDDs/PCDFs. For this reason, regulatory agencies established procedures for estimating risks associated with exposure to PCDDs/PCDFs, based on their potency relative to that of TCDD (12,69). The values developed by a workgroup of NATO are currently being used as the preferred toxicity equivalence factors by most international regulatory agencies. These international toxicity equivalence factors (I-TEFs) derived from a comparison of the relative potencies are presented in Table 3. To estimate the potency of complex PCDD or PCDF mixtures, the concentrations of the 2,3,7,8-chlorinated congeners in each isomeric group are multiplied by an appropriate I-TEF. For example, to derive a potency factor for a mixture of heptachlorodibenzodioxins (HpCDDs), the quantity of HpCDDs chlorinated at the 2,3,7 and 8 positions would be multiplied by a factor of 0.01 to obtain a $TCDD_{eq}$ (2,3,7,8-TCDD toxicity equivalent value). Thus, according to the toxicity ranking scheme, 2,3,7,8-chlorinated HpCDDs are considered to be 100 times less potent than TCDD.

Cleanup Criteria. The Centers for Disease Control (CDC; a branch of the U.S. Public Health Service) have established cleanup levels for 2,3,7,8-TCDD in soils in residential and industrial areas (70,71,72). These cleanup levels represent levels that the CDC feels do not pose a significant risk to people using these areas. For residential areas, a level of 1 ppb (ug/kg) has generally been proposed. For industrial sites, values of 5 - 50 ug/kg have been used, based on a consideration of the potential for exposure, whether or not children could be exposed, whether the TCDD will be covered with soil or asphalt, and the frequency with which high levels of dioxin were detected.

6.0 PERSPECTIVE

Regulatory toxicology as it currently is practiced involves selecting a single study (or at best a very small number of studies) from among a large and highly variable data set (including studies in humans, animals, microorganisms, etc.), and using this study in very simple models to arrive at a target criterion. The process of selecting the study and modelling the outcome incorporate a large number of complicated assumptions. Current approaches to developing criteria suffer from an over-attention to details of the model and the assumptions in the model while at the same time failing to use the full extent of available data. For example, in developing acute (single dose) toxicity criteria for TCDD and sodium fluoroacetate (compound 1080), study doses for both chemicals would be divided by a factor of 10 to account for interspecific variability. However, TCDD produces effects in different species over several orders of magnitude (2 μg/kg in guinea pigs through 5 mg/kg in hamsters) while sodium fluoroacetate causes acute toxicity in all animals, including man, at doses very close to 2 mg/kg (Range 1 - 15 mg/kg, with the exception of the opossum which has a LD_{50} of 0.1 mg/kg) (73). Several researchers are evaluating procedures to attempt to more fully use the database available for particular chemicals (36,74).

TABLE 3

International Toxicity Equivalency Factors (I-TEFs) and Proportion of Congeners of Concern in a Homologous Group

Congener of Concern	I-TEF	Congeners of Concern in a Homologous Group
PCDDs		
2,3,7,8-TCDD	1	1 out of 22 (5%)
1,2,3,7,8-PeCDD	0.5	1 out of 14 (7%)
1,2,3,4,7,8-HxCDD	0.1	3 out of 10 (30%)
1,2,3,4,7,8-HxCDD		
1,2,3,4,7,8-HxCDD		
1,2,3,4,6,7,8-HpCDD	0.01	1 out of 2 (50%)
OCDD	0.001	1 out of 1 (100%)
PCDFs		
2,3,7,8-TCDF	0.1	1 out of 38 (3%)
2,3,4,7,8-PeCDF	0.5	1 out of 28 (4%)
1,2,3,7,8-PeCDF	0.05	1 out of 28 (4%)
1,2,3,4,7,8-HxCDF		
1,2,3,7,8,9-HxCDF		
1,2,3,6,7,8-HxCDF	0.1	4 out of 16 (25%)
2,3,4,6,7,8-HxCDF		
1,2,3,4,7,8,9-HpCDF		
1,2,3,4,7,8,9-HpCDF	0.01	2 out of 4 (50%)
OCDF	0.001	1 out of 1 (100%)

Source: (69)

Hopefully, toxicity criteria that are still protective but that have a lower level of uncertainty can be developed.

For purposes of hazardous waste site risk assessment, in most cases the toxicity criteria developed by the agencies must be used. It is however, important to understand the extent of the uncertainty in these values, particularly because site regulatory personnel may not be aware of the conservatism that may be present. It is unlikely that a change in the toxicity value itself can be made for an individual site (although the EPA has allowed adjustments, it is a complicated process involving detailed review by the EPA headquarters staff). However, conservative assumptions used in the development of the toxicity criteria are an additional factor that should be considered in selecting the final remedy and target risk levels at a site.

7.0 GENERAL REFERENCES

Doull, J., Klaussen, D.C., and Amdur, M.O., *Casarett and Doull's Toxicology: The Basic Science of Poisons*. 2nd ed. Macmillan Publishing Co., New York. (1986)

Graham, J.D., Green, L.A., and Roberts, M.J., *In Search of Safety: Chemicals and Cancer Risk,* Harvard University Press, Cambridge, MA (1991)

Clayton, G.D. and Clayton, F.E., eds., *Patty's Industrial Hygiene and Toxicology.* Vol. 1: General Principles. 3rd. ed. Wiley-Interscience Publication, John Wiley and Sons, New York (1978)

Timbrell, J.A., *Principles of Biochemical Toxicology.* Taylor and Francis LTD., London (1982)

4.

EXPOSURE ASSESSMENT

An exposure assessment is conducted to estimate exposure to toxic chemicals from a hazardous waste site. The key components of the exposure assessment are:

- characterization of the potentially exposed populations,
- identification of actual or potential exposure pathways, and
- a determination of the extent of the exposure or potential exposure.

The first two components of the exposure assessment are designed to answer the questions: "Who is exposed?" and "How can exposure occur?" These two pieces constitute the qualitative portion of the exposure assessment.

Once the potential receptors and exposure routes have been identified, the final step in an exposure assessment is a determination of the potential amount of chemical that can be contacted via the various possible exposure routes (i.e., The question of "How much?" is addressed). Because of the uncertainty introduced by individual differences in behavior and physiology, and by the limited data on some of the parameters used in the exposure equation (e.g., limited data on how much soil people ingest or on how often a site is used), the final output of an exposure assessment is not actual exposure but only an exposure estimate. It is important to keep this distinction in mind in order to keep a focus on key issues and avoid overanalysis of minor factors. For example, worrying about whether people are exposed for 350 days/year (8) or 365 days/year (3), a difference in the exposure estimate of 4%, makes little sense when the overall assessment has an uncertainty of at least a factor of 10 (1000%).

Hazardous waste site exposure assessments typically do not answer the question: "What is the exposure that could occur at the site?" but in actuality, address the somewhat different question: "What is a reasonable maximum estimate of the exposure to individuals who might come in contact with the site?" The difference in these questions reflects the uncertainty in risk assessment and a desire on the part of regulators to ensure that public health is protected. In other words, hazardous waste site risk assessment is more focussed on protecting human health than in getting the right answer. The focus is somewhat misguided in that because of scarce resources (dollars and people), being overly conservative in the assessment of chemical hazards can increase risks from other causes. For example, limited funds can mean that dollars spent on hazardous waste investigation and cleanup are not spent on such things as health care. On the other hand, it is impossible to measure actual exposure (and risk), and as an affluent society we wish to err on the side of being health protective. However, limiting uncertainty to the extent possible in the exposure assessment should be a goal and will lead to maximizing the benefits of the risk assessment. At the very least, the policy decisions that are incorporated into the exposure assessment (i.e., the conservative assumptions) need to be noted in the risk assessment.

1.0 INTRODUCTION

A general principle of exposure assessment has been that conservative assumptions (i.e., assumptions that tend to overestimate exposure and therefore risk) are made at each step, so that the final estimates of exposure will be at the upper end of the likely range of actual exposures (the reasonable maximum exposure or RME approach) (3). However, the "standard" approaches and assumptions used to develop the RME estimate often produce estimates that are higher than this range (4). For example, while it is possible for a person to be in the lower 5th percentile in body weight, and it is possible for a person to be in the upper 95th percentile for water consumption, and it may be possible for this person to obtain all water from a well containing a key chemical at the 95% upper confidence limit of the average for the site, it is unlikely for all three to occur to the same individual. If the values could be known for certain, combining these three 95% upper exposure limits would yield an actual upper limit of $1\text{-}(1\text{-}.95)^3 = 99.9875$. In other words, the assessment would not be protective 95% of the time but 99.99% of the time. Further considering the other uncertainties (i.e., chemical concentrations in the well are unlikely to remain constant for the duration of exposure), exposure is clearly being substantially overestimated.

Recently, the EPA has suggested that risks and exposure be characterized more completely through the use of more than one risk descriptor (75). The EPA (75) suggests that both a "high end" and "central tendency" estimate of exposure and risk be presented. These terms and their definition are provided in Chapter 5 on Risk Characterization. In summary, the central tendency estimate is a value that attempts to indicate the general midpoint of the potential for exposure, while the high-end value is an estimate of the individual exposure for those persons at the upper end of the distribution. The EPA (75) also suggests that approaches to addressing uncertainty such as probabilistic risk assessment (Monte Carlo analysis) may be useful in more accurately characterizing risks.

This chapter presents the standard approaches to risk assessment as described in various EPA guidance documents, along with some of the commonly used assumptions. Enough information is provided to allow the reader to calculate exposure estimates for the key pathways at hazardous waste sites. However, providing these equations and assumptions is a secondary purpose of this chapter. The primary purpose is to provide some understanding of the numerous factors that need to be considered in preparing an exposure assessment and to give some feeling for the uncertainty associated with these exposure parameters.

2.0 RECEPTOR POPULATIONS

One important constituent of an exposure evaluation is a determination of the individuals (receptors) who may come in contact with chemicals from the contaminated area. Potential receptor populations for a hazardous waste site include all people in the area of the site who may be exposed to site chemicals either by direct contact on site or as a result of exposure to chemicals emanating from the site. The level of exposure experienced by different individuals is likely to vary considerably.

Workers and residents of a site, recreational users of the site, and workers and residents in adjacent off-site areas are the principal human populations potentially exposed.

In determining potential receptors, it is important to note people who may be particularly at risk. Considering these populations ensures the protection of other individuals who are less sensitive to chemical effects or that are less likely to be exposed. It may be useful to selectively consider other less sensitive populations only as necessary. Very young children are often the target population of greatest concern. This concern is for both scientific, objective reasons (small size means that the same chemical intake in mg/day yields a larger dose in mg chemical /kg bodyweight/day, and children ingest disproportionately high amounts of soil) as well as for more subjective reasons (parental concern). However, in many cases, children of a very young age are unlikely to enter a site. Children between the ages of about 8 and 12 probably represent the receptor population most likely to use sites with somewhat restricted access (either fenced or located some distance from residential areas) in a manner that would lead to contact with soil, sediments, and water.

Other receptor populations who may be particularly susceptible to chemical effects include both groups that are sensitive to the effects of the chemicals (e.g., the elderly are sensitive to respiratory toxicants) and people who for some reason may have increased contact with the site or with site chemicals. For example, people who subsist by fishing and farming may be more susceptible to increased levels of pesticides in fish than sportfishermen who catch and eat only a small number of fish each year. As another example, recent immigrants may use resources differently than we would expect, leading to the need to evaluate exposure routes that otherwise might not be considered.

3.0 EXPOSURE SCENARIOS

A key assumption that substantially influences the exposure estimate is the future use of the site. How people contact chemicals from a hazardous waste site and the amount of chemical that can be contacted are highly dependent on the use of the site and surrounding areas. For example, potentially huge differences in the likelihood of contacting chemicals in soils will exist between a site used for residential purposes (with seasonal landscaping activities and possibly a home garden) and a site used for a parking lot (contact only during construction). Therefore, estimating exposure requires an assessment of the type of land use possible at or near the site.

Ideally, a site would be monitored for an extended period of time, actual site uses and human behavior on the site documented, and this information used to arrive at appropriate exposure scenarios. In actuality, much more primitive approaches are used to estimate use. For example, a river running through a site may be used as a reason to include a fishing exposure scenario in the assessment. More definitive information can be collected from a site visit by looking for old fishing line and other fishermen's detritus along the river bank and by noting the state of the river. However, in many cases, the evaluation of exposure scenarios consists of educated guesses as to whether or not a particular type of exposure is occurring or could occur at the site.

The types of exposure that may be associated with current land use can be determined by simply observing activities at the site and in the surrounding areas. An assessment of exposure at a site under current use scenarios provides a snapshot of potential risks associated with the site as it currently exists. It is also important in determining whether to limit site access and in assessing if a site is a potential exposure source to nearby off-site residents.

If it is possible to determine that site use in the future will be the same as current use, the assessment of potential exposure is straightforward. However, in most cases, the potential site use will not be known with certainty or at least it is not possible to guarantee a particular future use. Because of this, it is generally necessary to evaluate potential exposure under various hypothetical future land use scenarios. As can be expected considering its influence on the results of the risk assessment, the choice of likely future use scenarios is a major source of debate. Land owners responsible for cleanup often argue that site use will remain consistent with current practices. On the other hand, regulators responsible for ensuring that public health is protected often want to evaluate "worst case" or at least maximum likely scenarios (in particular, residential use), given that there are no guarantees on land use and that they may be held accountable for future risks.

Two approaches can be used to address the problem caused by uncertainty in future land use. The first involves assessing a range of possible site use options. Scenarios for a site currently being used as an industrial area could include continued use for this same purpose, sale of the land and redevelopment for residential purposes, use of the land for commercial offices, or reuse of the land for a playground or park. Ideally, after evaluating the exposure and risks associated with these different scenarios, a cost-benefit analysis could be used to optimize future use of the land. The landowner, developer, and regulator could arrive at a mutually agreed upon use or uses for the land, with future different use requiring a re-evaluation of possible risks. Factors that could be considered in assessing future use would include value of the land for each of the possible future use options, cost of cleanup to allow each of these uses, and any other factors that might influence costs and desirability of the land (e.g., the need for deed restrictions).

What more commonly occurs is that even though a range of possible future use scenarios is evaluated, a single option, usually the most restrictive, is selected as the basis for site remediation goals. A reason for this is that even though all parties may want to see several options, without guarantees on future use, regulators whose job it is to protect human health, ultimately require maximum protection scenarios. Regulators may allow less restrictive scenarios if assured of no residential use, but such assurances may need to be in the form of deed restrictions. Owners typically do not want deed restrictions and tend to argue for less restrictive criteria without limits on use.

Another approach to addressing the issue of uncertain future use is to simply address the reasonable maximum use scenario and then only evaluate other scenarios if necessary. Exposure under a residential use scenario is likely to represent a maximum exposure condition. Therefore, evaluating this hypothetical exposure scenario will provide a good indication if exposure under other potential future use scenarios would be likely to pose health risks. For example, even if it may be more likely that a site would be reused for park land, any exposure associated with use for park land would be lower than that associated with residential use. If no risks are posed by the maximum future use scenario (i.e., if the cancer risks are estimated to be below the EPA's target risk levels and the estimated exposure is below reference doses for noncarcinogenic effects), then no further remedial action is warranted. However, if the residential use scenario may pose a risk, further evaluation of the more likely scenarios may be conducted. This evaluation may include a more detailed analysis of likely future uses of the area, additional investigative work, or an assessment of appropriate remedial alternatives.

4.0 EXPOSURE PATHWAYS

Once receptor populations have been identified and the potential exposure scenarios determined, the next step is to evaluate how people exposed under the different scenarios could contact site chemicals. As noted by the EPA (3), an exposure pathway consists of four main elements:

- A source and a mechanism of chemical release to the environment

- An environmental transport medium for the released chemical

- A point of actual or potential contact by a receptor with the contaminated medium (referred to as the exposure point)

- A route of intake (e.g., ingestion, inhalation, dermal absorption) at the exposure point.

If even one of these factors is absent, the pathway is incomplete and exposure cannot occur.

Determining appropriate exposure pathways to be considered in an assessment requires knowledge about the possible exposure scenarios, about the chemicals present and their behavior in the environment, and about characteristics of the site that may affect exposure, such as depth to groundwater and useability of this water. A list of some potential exposure pathways is provided in Table 1. A determination of all routes that may lead to exposure at the site is followed by an evaluation of the likelihood of exposure occurring via any of these pathways. As a first step, it may be useful to subdivide the exposure routes of potential concern. For example, the exposure that may occur on site can be separated from potential off-site exposures as a convenient method of separating areas of suspected chemical deposition (on-site areas) from areas likely to have lower contamination (off-site areas). As another approach, potential use of the site as it currently exists may be separated from potential future use scenarios. The particular approach used to subdivide exposure scenarios and in fact, the need to subdivide these scenarios is established on a site-by-site basis.

Depending on the chemicals at the site, site characteristics, and whether the exposure point is on site or off site, possible routes of exposure on a site containing hazardous wastes can include direct contact with contaminated environmental media (soil, sediments, and surface water) from outdoor activities (e.g., gardening or fishing), exposure through the domestic use of site ground water or surface water, exposure through the consumption of vegetables, home-raised livestock, or fish from the site, and inhalation exposure. Future use of a site for residential purposes would lead to exposure via these same routes, but the frequency and duration of potential contact with environmental media at the site would be greatly increased. For most hazardous waste sites, the source of contamination is usually site soils, and the nature of the chemicals involved greatly influence the potential for off-site migration and exposure.

TABLE 1

Potential Exposure Pathways

SOIL-BOUND CONTAMINANTS:
Direct ingestion of contaminated soil (either inadvertent or advertent)*
Dermal absorption following direct contact
Consumption of vegetables grown in contaminated soil
Consumption of livestock feeding in contaminated area
Migration of contaminants and exposure in other media (e.g., groundwater or air)

CONTAMINANTS IN GROUNDWATER:
Ingestion of drinking water*
Dermal absorption while bathing*
Inhalation of volatiles*
Consumption of vegetables watered with contaminated water
Consumption of livestock watered with contaminated water

CONTAMINANTS IN SURFACE WATER:
All pathways listed for groundwater may also be considered for surface water
Consumption of contaminated fish*
Dermal absorption while swimming
Ingestion while swimming

CONTAMINANTS IN AIR:
Direct inhalation*
Dermal absorption

---

*Exposure pathways that are most commonly critical.

In performing an exposure assessment, only routes of exposure that are considered possible at a site are generally discussed in detail. Exposure via other routes, if they occur at all, are likely to lead to much lower exposure than presented for these more probable routes. However, it is important to remember that actual exposure and perceived exposure can differ. Exposure pathways that may have sentimental concern should at least be considered. For example, children playing on a grassy site with groundwater contamination as the main concern may not be at risk, but the assessment should note reasons why exposure is unlikely, as parents will want this information.

5.0 QUANTITATIVE EXPOSURE ASSESSMENT

Individuals are exposed to materials released into the environment in varying quantities and proportions via a wide variety of exposure routes. The actual amount of material an individual is exposed to depends on the individual's frequency, extent, and duration of exposure, which in turn depends on many factors (e.g., location of residence, age, body weight, sex, and activity patterns). Patterns of exposure are highly variable among individuals. For example, the amount of soil ingested by a child may vary from less than 10 mg for a child playing on a grassy area or indoors in a very clean house, to relatively large amounts (10,000 mg or more) for children who are intentionally eating soil. This large potential variation in exposure to environmental contaminants is responsible for the uncertainty that is inherent in risk assessment.

5.1 Common Exposure Parameters

Exposure assessment can be summarized generically by the equation:

Eq (7)
$$CDI = \frac{CM \times IR \times EF \times ED}{BW \times AT}$$

Where:

CDI = Average daily intake of the chemical (mg/kg/day)

CS = Chemical concentration in the intake media (mg/g,L, or m^3)

IR = Intake rate for the media (g/day, L/day, or m^3/day)

EF = Frequency of exposure (days/365 days)

ED = Exposure duration (years)

BW = Body weight (kg)

AT = Averaging time (years)

Information on the intake levels for the different exposure media are described in the following sections. However, information on the duration of exposure, the frequency of exposure, body weights and the averaging time are generally similar for all routes.

Chemical Concentration. Chemical concentrations for use in exposure assessment ideally will be representative of the site and of site media that may be contacted. Chemical concentrations used in the exposure assessment are typically based on data collected as part of the remedial investigation. One problem with these data is that the focus of an RI (to determine the nature and extent of contamination) may not be consistent with the needs of the exposure assessment for samples representative of chemical concentrations in site media (76). For example, surface soil samples may consist of the upper foot or more of soil, while the actual soil that could be contacted (and therefore that is of concern in a risk assessment) is the top 1-2 cms. Using a foot thick sample, if a contaminant were all contained in the top inch of soil, would represent a dilution factor of approximately 10. This effect is balanced somewhat, however, because persistent chemicals are likely to mix into deeper soils over time and because of natural levels of metals contributing to the total levels of metals detected in a 1 foot sample. The risk assessor needs to carefully examine the data to ensure that it is representative of the site and of possible exposure at the site. The EPA (3) typically uses an upperbound estimate of possible concentrations (usually the 95% UCL of the average site concentration of a chemical) in order to ensure that exposure is not underestimated.

Exposure Duration. The exposure duration is simply the time period (in years) that an individual may be exposed to the chemical. The EPA (8) noted that the average length of time an individual in the United States remains at a single residence is 9 years and suggests that a value of 30 years (the upper 90th percentile) probably represents a reasonable maximum value for residential use. However, the potential for lifetime use of a house should be considered. For other scenarios, such as occupational exposure or current exposure to trespassers, shorter exposure durations are likely. For example, at a fenced site, it may be reasonable to assume that children between the ages of 8 and 18 (a 10 year exposure duration) are able and interested in trespassing on a site. Under this age, they may be unable to enter the site, and over this age they are unlikely to have an interest in entering the site. Selection of an appropriate exposure duration is generally a subjective process that requires making reasonable assumptions about possible site uses.

Exposure Frequency. In order to estimate likely maximum exposure levels, information or assumptions about the frequency of the exposure event occurring must be developed. As a default assumption, it is generally possible to assume daily exposure for such things as water intake, inhalation exposure etc., under the assumptions that certain individuals will be at home daily. This value has been modified somewhat, in that EPA (8) currently assumes exposure will occur for 350 days/year but as noted at the beginning of this chapter the difference between this value and 365 is immaterial.

Estimating potential exposure frequency on site can be somewhat more difficult, particularly if contact with site soils is of concern. Several authors have estimated the frequency of exposure to a site but no clear empirical data are available. Kimbrough et al. (72) assumed that people would only come in contact and ingest soil for 6 months each year. Schaum (77) noted that the ground is frozen for an average of 118 days/year in the coldest part of the United States, while in other parts, people can come in contact with outdoor soil year-round. He therefore determined that exposure could occur for anywhere between 247 and 365 days/year. Hawley (78) noted that people are likely to be exposed year-round to soil, both outdoors as soil, and indoors as a major component of dust. Sedman (79) reviewed a number of studies which examine relationships among lead concentrations in soil, house dust, and blood lead levels in young children. Correlations were observed among these parameters; however, in some studies, house-dust concentrations were higher than outdoor soils while in others the reverse was observed. In general, EPA assumes that soil ingestion will occur year-round, in residential dwellings, both from soil and that component of house dust that is soil.

In the case of people who infrequently visit a contaminated area, an estimate of the frequency of exposure should be derived for the particular site. Because it is generally not possible to determine the actual number of times individuals visit a site, assumptions are made that try to provide a reasonable but bounding estimate of potential frequency. One value that can be used is to assume that contact with site soils occurs 100 days/year (equal to both days of every weekend or every day of the summer). Other values that may be more appropriate in specific cases, depending on such factors as distance from the site to residential areas, fencing etc., include 50 days/year (once/week) and 190 days/year (half the year).

Body Weights. The EPA uses the average body weight in exposure calculations. The average adult weighs approximately 70 kg (140 lbs; 3,8). Children between the ages of 1 and 6 weigh approximately 15 kg and children between the ages of 6 and 11 weigh approximately 30 kg. Anderson et al. (8,80) list body weights for children at different ages and for adults and also list ranges of body weights for the different ages.

Averaging Time. For noncarcinogens, the averaging time is generally assumed to be equal to the exposure duration. However, for carcinogens, the dose is averaged over the lifetime of the individual (assumed to be 70 years) in order that the exposure dose be equivalent to the toxicity criteria. Cancer slope factors are calculated based on a full lifetime of exposure and consequently, exposure must be calculated in an equivalent manner.

5.2 Air

Inhalation exposure to volatile compounds or to chemicals bound to windborne particulates being released from a property can occur, particularly from sites that are not vegetated. Individuals on site and off-site residents may be exposed to chemicals that volatilize from hazardous waste site soils or that are released into the air from seeps. In most cases, waste disposal activities at a site will have ceased long ago, and any mobile compounds are likely to have already volatilized from surface soils. If these sites are left undisturbed, volatile emissions are unlikely but invasive activities can cause problems. For landfills, release of volatile chemicals can occur by migration through the surface soils, and particularly during the dry seasons, through cracks in the cover material and from any vents that open as a result of subsidence. Methane gas generated in the landfills can act as a carrier to increase the migration of other, more toxic volatile organic compounds through gas vents. During wet seasons some exposure may also occur due to release of volatiles from seeps at the base of the landfill.

Exposure to persistent chemicals such as metals may occur by inhalation of any dusts that are released into the air as a result of wind erosion of surface soils. The amount of a site that is covered by grass, buildings or pavement will greatly decrease the potential for airborne release. Surrounding tree cover or buildings can act to decrease wind velocity and consequently, to decrease the ability of the wind to entrain particulate matter. Soil moisture or snow cover also act to limit the potential for windborne migration of soils.

Volatile Emissions. Ideally, airborne concentrations of VOCs at the point of potential exposure (the exposure point) have been measured and are available for calculating exposure (81,82,83). More commonly, however, exposure must be estimated based on concentrations of chemicals measured in other media (typically soil) and the estimated air concentration developed based on modelling the release from this media. Inhalation exposure to site-related chemicals may occur as a result of migration of chemicals that volatilize from the site and are then transported by the wind or by diffusion to off-site locations. Inhalation exposure can also occur as a result of migration of chemicals upward through the vadose zone and into homes through any cracks in the foundation.

Contaminant release and air dispersion modelling can be used to estimate exposure to chemicals in air in the vicinity of a site. Exposure can be estimated using simple models (which tend to produce conservative estimates of exposure) and more complex models (EPA 1991). The more complex models take into account such site-specific factors as chemical and soil characteristics, windspeed, stability, the effects of terrain and structures on air flow, and distance to receptors. Air emission is not generally a critical pathway at hazardous waste sites. However, if this pathway is of concern, either simplified or detailed air modelling should be conducted.

As noted previously, volatile chemicals are unlikely to remain on the surface of a hazardous waste site but are likely to have either volatilized or to have leached downward towards groundwater. Consequently, assessing exposure to these compounds is generally unnecessary, but if required, it is a complicated process. However, metals and other persistent compounds that may remain on the site surface may migrate through wind dispersion of particulate matter. Screening level estimates of exposure to this airborne material can be calculated using simplified approaches.

Inhalation exposure to metals and other persistent organic compounds that are bound to soil particulate matter will depend on the amount of chemical that is present in the air and that is inhaled, and the amount of time the person is exposed (84). Ideally, information will be available on the amount of chemical or at least particulate matter actually airborne at a site. In many cases, however, estimates must be made. One approach used to estimate potential maximum exposure levels is to assume that the concentration of particulates in air is equal to the national ambient air quality standards for respirable particulate matter of 50 ug/m^3 or 5×10^{-8}kg/m^3, that all particulate matter in the air comes from the site, and that the concentration of chemical in airborne particulate matter is equal to on-site soil concentrations. EPA (9) in the RAGS Part B guidance manual has provided an alternative, somewhat less conservative value for the potential amount of particulate matter from a site. This value of 2.2×10^{-10}kg/m^3 at off-site locations was estimated using a modelling approach. The choice of values depends on needs and acceptability. In general, the EPA value should probably be used, with the more conservative value only used if the EPA number is unacceptable to regulators.

Once the exposure point concentration (in mg/m^3) has been estimated, the next step is to estimate dose. Because the size of an individual can influence the amount of a compound that can reach a target organ and cause an effect, health criteria are commonly reported in terms of a dose in units of milligrams of chemical/kilogram of body weight/day (mg/kg/day). Standard assumptions used in risk assessment are that the average person weighs 70 kg and breathes 20 m^3 of air/day (3). Because breathing rate is directly correlated with size, no specific adjustment for exposure in children is made, although children may receive slightly greater exposure per unit body weight than adults.

For noncarcinogens, exposure is averaged over the duration of exposure while for carcinogens, the averaging time is expressed in terms of a full lifetime of 70 years (3). Using these assumptions, exposure can be estimated using the formula:

Eq (8) $$EXP = \frac{AC \times BR \times EF \times ED}{BW \times AT}$$

where, EXP = Inhalation intake (mg/kg/day)

AC = Concentration of chemical in air (mg/m^3)

BR = Breathing rate (20 m^3/day)

EF = Exposure frequency (365 days/365 days)

ED = Exposure duration (30 years)

BW = Body weight (70 kg)

AT = Averaging time (30 or 70 years)

5.3 Water

Use of groundwater or surface water that has been affected by a site as a source of domestic water can lead to exposure via three pathways:

ingestion, dermal absorption, and inhalation (volatiles only). In addition to these pathways, exposure to chemicals in surface water can also occur during recreational use of the water and through the consumption of fish obtained from a contaminated source.

Groundwater. Exposure assessment for groundwater contamination generally focuses on the ingestion pathway. However, several studies have been completed that address exposure to volatiles via dermal absorption while bathing or via inhalation while showering or using domestic water for non-contact purposes such as washing dishes, in-home clothes washing, and flushing toilets (27,85,86,87,88). Both inhalation and dermal absorption appear to contribute doses of volatile compounds that are roughly equivalent to the dose delivered by ingesting water at the standard rate of 2 liters/day (27). However, a considerable amount of uncertainty is associated with calculating doses received via these pathways. One approach that can be used to account for exposure via these other pathways is to incorporate a factor of two into the dose estimates developed based on ingestion exposure. Considering that conservative exposure parameters are used in calculating exposure via ingestion, inhalation, and dermal absorption, and that the combination of these parameters could lead to very conservative estimates of risk, the use of a factor of two rather than a factor of three seems appropriate.

Surface Water. If surface water can be used as a source of domestic water, users would be exposed to any chemicals in the water via the same pathways as for domestic use of groundwater, namely, ingestion, inhalation, and dermal absorption. Other routes include recreational exposure while swimming or exposure through the consumption of fish. Swimming or wading can lead to exposure to chemicals present in surface water through dermal absorption or incidental ingestion. However, exposure is very likely to be of short duration and frequency and this exposure route is seldom likely to pose substantial risks at a site. Fish can bioaccumulate certain chemicals, including mercury and high molecular weight organochlorine compounds such as PCBs and DDT, and consumption of fish from chemical-containing water could be of concern (fish consumption is discussed in Section 5.5).

Dose Calculations - Domestic Water. Assumptions for drinking water exposure include the assumption that an adult drinks 2 liters of liquid/day. The EPA (8) notes that in most cases, this liquid is not strictly water and not from a single source. However, as a generally conservative estimate, 2 liters of water is a reasonable value. Other assumptions are the same as noted for the inhalation pathways, with the exception of relative absorption fraction.

The relative fraction absorbed is the percent of chemical absorbed from water relative to the percent of chemical absorbed from the medium used in the animal study that served as the basis for the toxicity criterion. For example, if 60% of a chemical dose is absorbed from water and 75% is absorbed from corn oil in a chronic animal bioassay, the relative fraction absorbed is 60%/75% = 80%. For chemicals in water, a relative absorption default value of one is often used if information on actual absorption is not readily available.

Exposure via ingestion of groundwater can be calculated using the equation:

Eq (9) $$CDI = \frac{CW \times IR \times CF \times FI \times EF \times ED}{BW \times AT}$$

Where:

CDI = Average daily intake of the chemical (mg/kg/day)

CW = Chemical concentration in water (ug/L)

IR = Water consumption rate (2 L/day)

CF = Conversion factor (10^{-3}mg/ug)

FI = Relative fraction absorbed from water (100%)

EF = Frequency of site contact (350 days/365 days)

ED = Exposure duration (30 years)

BW = Body weight (70 kg)

AT = Averaging time (30 or 70 years)

Using these equations and an estimate of the representative chemical concentration in groundwater, chronic daily intake levels can be calculated for groundwater ingestion. As noted previously, for volatile and semi-volatile compounds, an adjustment factor of two is multiplied by the estimate to ensure that exposure via dermal absorption and inhalation is considered. If more detailed evaluation of these pathways are desired, literature by the EPA (27) McKone (88), and Foster and Chrostowski (86) are good starting points.

5.4 Soils and Sediments

Direct contact with on-site surface or exposed subsurface soils may lead to exposure via dermal absorption of chemicals present in the soils or via inadvertent or intentional ingestion of on-site soils. Direct contact with sediments in streams, ponds or marshes or with contaminated surface water can also lead to dermal absorption or ingestion exposure. Sediments can also act as a source for chemicals to be taken up by aquatic organisms and ultimately by fish that are consumed by humans (fish consumption is discussed in section 5.5).

Soil. Only non-mobile chemicals such as metals and persistent organic compounds are generally considered to pose a risk for direct contact exposure to surface soils. Mobile chemicals will tend to either volatilize from or percolate downward through the soil surface, which is the layer that people are likely to contact. Short-term exposure to volatile or semi-volatile compounds in soils is possible during activities that involve digging into soils, but even in these cases, volatilization of the chemicals from the newly exposed soils is likely to occur in a relatively short time period (days to weeks). Considering that most risk assessments are designed to evaluate risks over extended periods (months to years), volatilization and leaching generally ensure that volatile and semivolatile compounds do not pose direct contact risks.

Exposure via direct ingestion of soils is of greatest concern if the soils are in an area that is frequently used (as in a backyard) and if the soils are not covered by grass, pavement, or buildings. On-site workers engaged in digging or earth moving activities and local children playing on a site are most likely to contact soils. Very few sites have large areas of bare soil that remain in that state for a long period of time. In most cases, surface soils are covered with pavement or vegetation, and the presence of a substantial cover is likely to limit exposure. Metals and non-mobile persistent organics deposited at a site are unlikely to migrate upward through the cover material. Consequently, unless the surface soils are disturbed or erosion exposes subsurface soils, direct contact with the surface material covering most of the site is unlikely to be a significant route of exposure. The presence of some form of cover probably also limits exposure in residential settings. However, in certain environments (such as urban play areas, frequently used areas where soils will be constantly disturbed, and unvegetated arid sections of the country) exposure by direct contact is a particular concern.

Any on-site construction activities, such as building houses, would lead to disturbing subsurface soils, possibly to a significant depth, and the mixing of surface and subsurface soils. Construction workers can clearly be exposed to these soils by direct contact. Subsurface soils are generally nutrient-poor and are not used for cover soils. Consequently, exposure to chemicals in subsurface soils will be limited in most cases.

Sediments. Children or adults using or playing in swampy areas may be exposed by direct contact with sediments. In general, it is unlikely that such exposure would occur frequently or for any substantial duration and consequently, this route may not pose substantial risks. However, it is often included for completeness.

Dose Calculations - Ingestion. Exposure to chemicals in soil can occur via direct contact with the soil and subsequent inadvertent ingestion associated with hand-to-mouth contact. Exposure can also occur via direct ingestion of contaminated soil. The direct ingestion of nonfood items is termed pica; soil-specific pica is termed geophagia. Pica is a rather general term that includes both the normal experimental ingestion of nonfood items common in children and the abnormal and relatively uncommon craving for, and habitual ingestion of, nonfood items. This latter is generally associated with medical conditions such as malnutrition, certain neurobehavioral disorders, iron deficiency anemia, and occasionally with a particular cultural background, and can be termed habitual pica (89). Because habitual pica is rare, it generally is not considered in risk assessment unless conditions in a particular area suggest that susceptible individuals may be present. Pica, the ingestion of nonfood items that is common in young children, is considered in risk assessment.

The residential use scenario is generally considered likely to lead to the greatest exposure via soil ingestion. Landscaping or gardening may lead to greater and more regular contact with soils than would occur in nonresidential areas. Residential use also assumes that the potentially exposed individuals can include very young children who ingest substantially higher amounts of soil than older children and adults. Because young children ingest more soil but weigh less than adults and older children, they have an increased potential for significant exposure via this pathway. Consequently, it may be important to evaluate this pathway separately for children, particularly for noncarcinogenic risks and for exposure to acutely toxic agents.

For carcinogenic effects, where exposure is averaged over a lifetime rather than over the exposure period, both childhood and adult exposure is considered. The standard approach used by the EPA is to assume that an individual resides in the same location for 30 years and ingests soil at 200 mg/day and weighs an average of 15 kg for the 5 years between the ages of 1-6, ingests 100 mg/day of soil and weighs 30 kg between the ages of 6 and 11, and ingests 100 mg/day of soil and weighs 70 kg between the age of 11 and 30. It is further assumed that the individual is exposed 350 days/year to site soils and has an average lifespan of 70 years (3,8).

Contaminants present in soils may not be as bioavailable in the human body as contaminants present in other environmental media such as water or food (90,91). In order to account for this difference in bioavailability, the ratio of the amount of contaminant that can be absorbed in the body from soil to the amount that can be absorbed from other media is factored into the equation used to calculate the CDI value.

For noncarcinogens, exposure is averaged over the duration of exposure, while for carcinogens, the averaging time is 70 years (3,8). Using these assumptions, exposure can be estimated using the formula:

Eq (10) $$CDI = \frac{CS \times IR \times CF \times FI \times EF \times ED}{BW \times AT}$$

Where,

CDI = Average daily intake of the chemical (mg/kg/day)

CS = Chemical concentration in soil (mg/kg)

IR = Soil ingestion rate (mg/day)

CF = Conversion factor (10^{-6}kg/mg)

FI = Relative fraction absorbed from soil (%)

EF = Frequency of site contact (days/365 days)

ED = Exposure duration (years)

BW = Body weight (kg)

AT = Averaging time (years)

Using this equation and a representative soil concentration for surface soil (ideally the upper inch and preferably less than the upper 6 inches of soil), chronic daily intake levels can be calculated.

Dose Calculations - Dermal Absorption. Children and young adults using a site may be exposed to any chemicals in soils or sediments via dermal absorption of chemical from soils adhering to body surfaces. The duration of exposure may be longer than the period of contact with the site, because site users may get dirt on their bodies and not wash it off for some time. For most assessments, a conservative assumption is made that the site is the source of all soil that the person encounters on a given day.

The EPA (27) has developed a range of recommended values for dermal exposure factors. These values are not necessarily accurate but do represent a starting point for exposure calculations. For incidental dermal exposure, EPA assumes that 5000 cm^2 is indicative of a central tendency value and that 5800 cm^2 is a reasonable upper value. Both these values imply that the entire area of the head, forearms, hands, and lower legs are exposed. Another value that is used is 2000 cm^2, which is approximately equal to the surface area of the hands, feet, and lower arms (or hands, feet and lower legs) of 10 year old children (80). The EPA (27) presented central tendency and upper estimates of dermal contact rates of 0.2 and 1.0 mg/cm^2. EPA also noted that dermal contact rates may be lower on areas of the body other than the hands.

For purposes of assessing dermal risks, the EPA typically calculates the amount of chemical absorbed into the body. The absorbed dose must then be compared with an equivalent toxicity value to evaluate the potential risks. However, most toxicity criteria are expressed in terms of dose at the point of entry into the body (i.e., the mouth for oral exposure) not in terms of absorbed dose. Consequently, the toxicity value must be adjusted to account for oral absorption or for some chemicals that are almost completely absorbed, it is assumed that the oral and absorbed toxicity criteria are equivalent.

Another approach is to base the calculations on the relative percent of a chemical absorbed into the body dermally versus orally. As a default, it can be assumed that 1 percent as much of the inorganic materials and 10 percent as much of organics will be absorbed through the skin as from the gastrointestinal tract (90,92,93). These values are almost certainly conservative (health protective). The primary function of the gastrointestinal tract is to allow the absorption of both minerals (i.e., metals) and nutrients (i.e., organic materials) into the body. The primary function of the skin, however, is to act as a barrier to entry of foreign materials. Consequently, absorption of materials from the gastrointestinal tract following oral exposure is generally considered to occur much more readily than dermal absorption. In addition, once a material is ingested, it will remain in contact with the gastrointestinal tract for approximately 24 hours or more, while materials on skin will most likely be washed off more frequently. An advantage of this approach is that oral toxicity criteria can be used, under the assumption that dermal absorption of organics is 10 percent of oral absorption and metals are 1 percent as available from the skin as from the gastrointestinal tract. In other words, the CDI for dermal absorption is expressed relative to oral absorption, allowing the oral RfDs and SFs to be used in calculations. As noted above, this approach differs somewhat from common practice, but is mathematically equivalent and provides a better reflection of our actual state of knowledge on absorption.

For noncarcinogens, exposure is averaged over the duration of exposure (9 years - central tendency; 30 years - upper value). For carcinogens, cancer potency factors are developed based on a full lifetime exposure and therefore the averaging time is also expressed in terms of a full lifetime of 70 years (EPA 1989). Using these assumptions, exposure can be estimated using the formula:

Eq (11)
$$CDI = \frac{CS \times SA \times CR \times CF \times AD \times EF \times ED}{BW \times AT}$$

Where:

CDI = Average daily intake of the chemical (mg/kg/day)

CS = Chemical concentration in soil (mg/kg)

SA = Skin surface area contacted (2,000 cm^2/day)

CR = Soil-to-skin contact rate (1 mg/cm^2)

CF = Conversion factor(10^{-6}kg/mg)

AD = Relative fraction absorbed (1% or 10%)

EF = Frequency of site contact (100 days/365 days)

ED = Exposure duration (5 years)

BW = Body weight (30 kg)

AT = Averaging time (5 or 70 years).

Chronic daily intake levels can be calculated using this equation and a representative soil concentration for preferably the upper 6 inches or less of soils.

5.4 Food

Chemicals on or in food are an obvious source of exposure. However, a great deal of uncertainty is associated with evaluating this pathway and a large amount of detail is required for the assessment.

Consumption of Local Produce and Livestock. The amount of exposure that an individual receives through the ingestion of produce (fruits, vegetables, and grains) raised in the area near a site or even on a site depends on a large number of factors, including the deposition rate of particulates on to the produce, the concentration of chemicals in the soil, the form of the chemicals in the soil, uptake of chemicals by the plant, and the amount of locally grown produce consumed by an individual. The amount of exposure that an individual would receive through the ingestion of meat raised on or near a site would also depend on a large number of factors including the amount of meat consumed by the individual each day, the number of times locally raised meat was consumed per year, pharmacokinetics of the particular chemical in the livestock, the amount of locally grown feed, soil, and water consumed by the cattle, the amount of contamination in the feed, soil, and water, and the uptake of the chemical by feed plants. For produce, metals which can be taken up by the plants are of particular concern. In meat, chemicals that tend to bioconcentrate, such as PCBs and PCDDs/PCDFs, are considered most likely to pose a health risk. Soil ingestion by livestock and subsequent incorporation of chemicals into the body can be an important exposure route for chemicals that are not translocated by plants.

The amount of exposure that an individual is likely to receive via ingesting locally grown foodstuffs is highly dependent on site specific factors. As noted by McKone and Ryan (94) there is a very large (two orders of magnitude) amount of uncertainty associated with this pathway. Consequently, this pathway was not evaluated in detail in this chapter. The EPA (7) details equations and assumptions used in estimating chemical intake through food.

Fish Consumption. Material that is emitted from a site and subsequently enters surface water bodies can be ingested by fish. The amount of chemical exposure associated with eating fish is dependent on the amount of material that enters the water, chemical characteristics of the body of water, physical characteristics of the body of water, the amount of the chemical that is taken up by the fish, the bioconcentration rate of the chemical, and the amount of fish from the particular body of water that is eaten on a regular basis. Because most relatively mobile materials (i.e., chemicals likely to move to waterbodies) do not bioconcentrate in fish, ingestion of fish is unlikely to pose a serious risk for most chemicals. However, for chemicals such as PCBs, PCDDs/PCDFs, mercury, and organochlorine pesticides that are bioconcentrated, this exposure route can be of concern. A compound that is noteworthy for being less of a concern in fish than would be expected is arsenic. Arsenic is stored by fish in a methylated form, but arsenic in this form appears to be unavailable to humans (95). Finally, PAHs would be expected to bioconcentrate in fish based on their high K_{oc} values but because these compounds are rapidly metabolized and excreted they tend not to be present in fish. Mollusks such as clams, oysters, and scallops, however, are an exception in that these organisms are unable to metabolize the PAHs.

Chemicals in sediments can be taken up by benthic organisms and can be ingested by predatory fish using the benthic organisms as a food source (96). Chemicals in sediments can also be released slowly into the water column and can be taken in by any organisms in the affected water body. Certain chemicals can biomagnify in the ecosystem, and animals in higher trophic levels such as top predators are likely to have the highest levels of these chemicals. As an example, organochlorine pesticide levels are likely to be non-detectable in surface water, barely detectable in small benthic organisms, present at slightly higher levels in small fish, and present at the highest levels in top game fish such as pickerel, trout, and bass. Consumption of fish from areas containing elevated levels of chemicals that biomagnify is complicated by the fact that humans tend to consume gamefish, which also happen to be top predators.

Dose Calculations. Actual exposure to individuals from the ingestion of fish probably varies considerably and for this reason is difficult to assess accurately. For example, the fish ingestion rate used by the EPA in several of its analyses is 6.5 g/day (approximately 5 pounds/year) as an annual average (97). Many avid fishermen are likely to eat considerably more than this amount of fish each year, possibly from a single source if the lake or pond is large enough. The EPA (7) in its Exposure Factors Manual, averaged 90th percentile upper limits from two studies of 54 g/day and 225 g/day to arrive at a value of 140 g/day for large water bodies. Burmaster et al (98) evaluated the distribution of fish consumption by adults in Ontario, Canada, for use in probabilistic risk assessment (Monte Carlo analysis). For fishermen who consumed fish, the average intake was 20 g/day, the median intake was 10 g/day and 95 percent of the fishermen consumed less than 75 g/day. Based on the information in these studies, a bounding estimate is probably best provided by a value of 100 g/day. This number probably also best reflects the uncertainty in the estimate.

The EPA (7) noted that the available data were too limited to allow determination of values for smaller bodies of water. In addition, small bodies of water may not have sufficient fish to allow regular consumption. To partially address this issue, the fraction of the fish ingested that are likely to come from the water body of concern is often added to the exposure dose calculation. Better yet, it may be possible to collect site-specific information that can serve as a guide to understanding fish consumption habits in a particular region. For example, creel surveys for particular lakes may be available from local Fish and Wildlife Departments.

The exposure via ingestion of fish can be estimated using the formula:

Eq (12)

$$CDI = \frac{CF \times IR \times FI \times EF \times ED}{BW \times AT \times 10^3\ (g/kg)}$$

Where,

CDI = Average daily intake of the chemical (mg/kg/day)

CF = Chemical concentration in fish (mg/kg)

IR = Fish ingestion rate (g/day)

FI = Fraction from target water body (20% is a default)

EF = Frequency of ingestion (daily IR so 365 days/365 days)

ED = Exposure duration (years)

BW = Body weight (kg)

AT = Averaging time (years)

Chronic daily intake levels can be calculated using this equation and a measured or estimated chemical concentration in fish.

5.5 Occupational Exposure

Worker exposure can occur via three primary routes: inhalation, ingestion, or dermal absorption. Volatile chemicals or chemicals that become airborne as a component of windborne dust particles can be inhaled by workers, particularly in dusty areas. Workers can be exposed by ingestion as a result of normal hand-to-mouth activity (such as smoking, wiping one's mouth, or eating) or via ingestion of food that becomes contaminated in the workplace (e.g., by placing the food on a dusty surface, by accidentally dropping food on the floor, or by handling the food with dusty hands). Finally, dermal absorption can occur if workers get contaminated dust on exposed parts of their body and then absorb the material through their skin. Individual exposure and risk is highly variable depending on individual habits and in most cases defining the true exposure is impossible.

For workers, EPA, OSHA, and other agencies commonly use the average human body weight of 70 kg. The EPA (8) has determined that a soil ingestion rate of 50 mg/day should be used as a standard default value for worker exposure. However, the EPA notes that a higher soil ingestion rate of 480 mg/day (500 mg/day more accurately reflects the uncertainty in this value) may need to be used for certain industrial activities that involve more frequent contact with soils.

For dermal exposure, it is necessary to estimate the amount of dust or soil that gets on an area of skin, the amount of skin that is uncovered and therefore can have dust on it, and the amount of chemical that can be absorbed through the skin. The amount of dust or soil on skin has been estimated by various researchers, with values ranging from 0.2 to 1 mg/cm^2 of exposed skin (27). Schaum (77) estimated that a worker in a warm climate wearing a short sleeved shirt and long pants would have 2940 cm^2 of skin exposed. This value is probably reasonable for a worker in a dusty factory but the value of 2940 cm^2 is too precise and a value of 3000 cm^2 should be used.

6.0 UNCERTAINTY IN EXPOSURE ASSESSMENT

Estimates of exposure developed using the approaches and assumptions outlined in this chapter (and in EPA guidance) are conservative in two independent ways. First, for each exposure scenario, the assessment is focused on the locations where exposure is expected to be greatest, and on the subpopulations likely to be most exposed (for example, workers or children who frequent the most contaminated areas of the facility). For this reason, the estimates of exposure are really only valid for the most highly exposed persons, referred to as the maximally exposed individual (MEI) and will overestimate exposure for most people. Second, exposure assessments for these persons incorporate health protective assumptions about many factors that influence exposure, including emission rates, deposition rates, environmental persistence, bio-uptake, bioavailability, and human activity. For this reason, the exposure estimates are intended to overestimate exposures even for the MEI.

This "standard" approach is appropriate for developing a bounding estimate of exposure and ultimately of risk. That is, the approach provides an estimate that is at one end of the range of possible answers to the question of how much exposure could occur. However, if a more accurate estimation of exposure is required to allow better balancing of costs and benefits, the bounding estimate approach may not be appropriate. Approaches used to address the need for increased accuracy have included developing multiple estimates (best estimate and reasonable maximum estimate; more recently revised as the central tendency and upper value estimates) and considering both in the decision making process. A problem has been that often only a single value (invariably the upper estimate) is ultimately used in the decision. Recently, increasing support is being given to a probabilistic approach to developing estimates (4). The Monte Carlo approach has the advantage of being based on a distribution of the entire data set rather than relying on point estimates of specific parameters.

7.0 PERSPECTIVE

Exposure assessments are commonly designed to provide conservative estimates of likely exposure based on a desire to protect human health. However, as noted by Burmaster and Lehr (4), the final estimate of risk in some cases is so overly conservative as to be unrealistic. Because of the substantial impact that overspending on hazardous waste site remediation can have on the allocation of scarce resources, it is critical that reasonable estimates of exposure be used in the risk assessment.

8.0 GENERAL REFERENCES

Environmental Protection Agency (EPA), Risk Assessment Guidance for Superfund: Human Health Evaluation Manual, Part A, Interim Final. Office of Solid Waste and Emergency Response, Washington, DC, OSWER Directive 9285.701A, July (1989)

Environmental Protection Agency (EPA), Exposure Factors Handbook. Office of Health and Environmental Assessment, Washington, DC, EPA 600/8-89/043, July (1989)

Environmental Protection Agency (EPA), Superfund Exposure Assessment Manual. Office of Remedial Response, USEPA, Washington, DC, EPA 540/1-88-001, April (1988)

Environmental Protection Agency (EPA), Human Health Evaluation Manual, Supplemental Guidance: "Standard Default Exposure Factors." Office of Solid Waste and Emergency Response, Washington, D.C., March 1991. OSWER Directive 9285.6-03 (1991)

Environmental Protection Agency (EPA), Dermal Exposure Assessment: Principles and Applications (Interim Report) Office of Research and Development, U.S. EPA January 1992, EPA/600/8-91/011B (1992)

5.

RISK CHARACTERIZATION

..."The qualitative and quantitative risk characterization is only one of many important factors that must be considered in reaching the final decision [on remedial action at a hazardous waste site]" (75).

It is important to remember that because of uncertainty and variability in individuals' behavior and response to chemicals, it is not possible to determine the actual risk at a hazardous waste site. Consequently, while it is valuable to develop a numeric estimate of risk, it is at least as important to describe the non-quantifiable factors that are likely to affect the potential for risks to be posed by a site. Perhaps the most common mistake in the use of risk assessment in remediation is an over-reliance on a numeric solution to problems. A critical function of this section of a risk assessment is to clearly inform the reader/user that the quantitative estimate of risk is just one of many factors that need to be considered in reaching a remedial solution at a hazardous waste site.

1.0 CALCULATIONS OF SITE RISKS

The calculations required for the risk characterization section are generally quite simple. Information on the potential levels of exposure to contaminants (from the Exposure Assessment) is combined with information on the toxicity of the contaminants (from the Hazard

Evaluation) in order to calculate the potential health risks to individuals living near a hazardous waste site:

RISK = EXPOSURE x TOXICITY

In general, approaches have been used that estimate effects based on an administered dose of a chemical (i.e., how much contacts the body). However, more elaborate approaches that measure effects based on levels present in the body (the body burden) are being used more frequently.

1.1 Standard Calculations

The basic assumptions underlying cancer risk assessment differ somewhat from the assumptions underlying the evaluation of non-cancer effects, and consequently, these effects are calculated somewhat differently. The cancer risk is obtained by multiplying the dose of the chemical expressed in terms of the chronic daily intake (CDI; from the exposure assessment) by its cancer potency factor or slope factor (from the toxicity section):

Eq (13) $$\begin{aligned} \text{Risk} &= \text{Dose x Cancer Slope Factor} \\ &= \text{CDI [mg/kg/day] x CSF } [(\text{mg/kg/day})^{-1}] \end{aligned}$$ [1]

The risk value, which denotes the probability of getting cancer under the conditions presented in the assessment, can then compared with target risks levels established by regulatory agencies to determine if the risk is "significant".

To evaluate possible risk from exposure to noncarcinogenic contaminants, the CDI is divided by the health criterion value (the

[1]This equation, the linear low-dose model, is appropriate for risks below 10^{-2} but for higher risks an alternative model, the one-hit model, which is expressed as risk = 1-exp(-CDI x CSF) should be used. However the results from the two models are virtually the same below a risk level of approximately 10^{-2}. Considering that any risk exceeding 10^{-2} is clearly a problem (i.e., is well above regulatory criteria), the simplified form is generally acceptable. The obvious constraint on use of the simple model is that probabilities cannot exceed one and in general, risks exceeding 10^{-2} should probably be noted as undefined but greater than this value (i.e., $>10^{-2}$).

reference dose or RfD) to give the hazard quotient (HQ):

Eq (14) $$\begin{aligned} HQ &= \text{Exposure Dose/Reference Dose} \\ &= \text{CDI(mg/kg/day)/RfD(mg/kg/day)} \end{aligned}$$

If the CDI:RfD ratio (the HQ) is less than 1 (i.e., if the daily intake or dose is below the health criterion), the contaminant is considered unlikely to pose a health hazard to individuals exposed under the given scenario (3).

EPA typically expresses cancer risk in terms of an upper-bound excess lifetime cancer risk level. The term excess cancer risk indicates that the estimated cancer incidence is in addition to the normal background incidence of cancer in the human population. The term upper-bound indicates that the cancer slope factor is generally not the best estimate of the slope of the line relating cancer incidence to dose but is instead the upper 95% percent confidence limit of this slope (3). Furthermore, as noted in the toxicity chapter of this book, EPA derives slope factors and RfDs using a series of health protective assumptions, so that actual risks associated with exposure to a potential carcinogen are unlikely to exceed the risks estimated using these slope factors, but may be lower.

A general principle of risk assessment as practiced by EPA and other regulatory agencies has been that conservative assumptions (i.e., assumptions that tend to overestimate exposure, toxicity, and therefore risk) are made at each step, so that the final estimates of risk will be higher (and often much higher) than the upper end of the likely range of actual risks (3,12). More recently, EPA has suggested that risks be characterized more completely through the use of more than one risk descriptor (75). The EPA (75) suggests that both a "high end" and "central tendency" estimate of risk be presented. The "high end" estimate is defined as follows:

> The high end risk descriptor is a plausible estimate of the individual risk for those persons at the upper end of the risk distribution. The intent of this descriptor is to convey an estimate of the risk in the upper end of the distribution, but to avoid estimates which are beyond the true distribution. Conceptually, high-end risk means risks above about the 90th percentile of the population distribution, but not higher than the individual in the

population who has the highest risk (75).

The central tendency is defined as follows:

> The risk descriptor addressing central tendency may be either the arithmetic mean risk (Average Estimate) or the median risk (Median Estimate), either of which should be clearly labeled. Where both the arithmetic mean and the median are available but they differ substantially, it is helpful to present both (75).

The EPA notes that the purpose of providing both "high end" and "central tendency" estimates is to provide risk managers with a better characterization of risks and the uncertainties in order to facilitate making informed decisions at sites. As noted by EPA, the problem with the RME approach was that site decisions were often based on the single numeric value, without consideration of the complexities of the risk estimate. Use of an approach that provides several risk descriptors, including individual risk estimates (both "high end" and "central tendency"), population risk estimates, and estimates for different subgroups of the population may suggest to a risk manager that several complex issues need to be considered in making a risk-based decision. EPA (75) notes that probabilistic estimates of risk (Monte Carlo-type analyses) can also be useful in more accurately quantifying the "high end" and "central tendency" values. The results of a Monte Carlo analysis may also provide a better picture of the numeric risks at the site.

Although the more recent EPA (75) approach to risk assessment conceptually represents a substantial improvement on the risk characterization process from the RME approach, there are several potential problems. The Superfund Public Health Evaluation Manual (SPHEM; 6), EPA's first Superfund risk assessment guidance document, suggested that both upper-bound and average (central tendency) estimates of risk be provided. However, even though both values were presented, the average risk value was commonly ignored in making risk management decisions at sites and in deriving target cleanup values for environmental media. This tendency to disregard all but one value was partially responsible for the development of the EPA's 1989 policy of using only a single RME value (3). It is possible that increased education of regulatory risk managers will increase the usefulness of multiple risk values, but it seems just as likely that the single "high end" risk number will continue to be used as the numeric benchmark for regulatory decisions. This approach is still likely to be less conservative than the RME approach because the high-end value is designed to approximate the 90-percent UCL on risk rather than a bounding estimate. However, the disadvantage of using this high-end value is that the public (and probably most health practitioners) may not be particularly satisfied with an estimate that suggests that around 90 percent of the individuals are unlikely to be at significant risk and conversely, that 10 percent may be exposed to levels of potential concern. A bounding estimate of risk (i.e., an assessment designed to protect all or almost all individuals by probably overestimating risks) appears likely to be somewhat more acceptable to the general population.

The EPA (75) notes that there are different approaches available for estimating a "high end" exposure value, with the choice among different approaches dependent on the amount of information available on the distributions of the exposure parameters. For virtually all risk assessments, only a limited amount of information is actually available on the distributions of at least some of the parameters. Therefore, an approach that consists of developing a bounding estimate and then "backing off" this value until reaching a value in the "high end" range may be the best approach to addressing both uncertainty and the need for practicality in the values. The amount of "backing off" will be dependent on such factors as necessity (if no risks are posed by the high-end value then no backing off is required), costs of remediation, public concern, etc.

Because of the uncertainty inherent in the risk assessment process, an overriding principle is that the assumptions, approaches, and criteria used in the risk assessment should be clearly and logically described. Clearly, documenting the procedures used in the risk assessment allows the verification of all calculations and provides perspective regarding the degree of conservatism inherent in the approaches and assumptions used.

1.2 Calculations Based on Body Burden

Chemicals detected at similar concentrations in environmental media having different characteristics can have quite different effects on the exposed individual. For example, Poiger and Schlatter (90) noted that TCDD freshly mixed with carbon was much more bioavailable than the same level of TCDD allowed to remain in contact with the carbon for several hours. As another example, metals may be present in the environment in a number of different compounds and the influence of the associated elements on bioavailability of the metal can be substantial (11).

Measuring the binding potential and leachability of site-specific chemicals is one approach that can be used to better estimate exposure. However, the best approach (from a technical standpoint) is to measure the chemical in the body and preferably in the target organ of concern (99). An obvious problem (i.e., the reason why this approach is not simply the best approach and why it is not always feasible) is that to measure a chemical in the body, exposure has to have occurred. For hazardous waste sites, most evaluations are intended to prevent or at least severely limit exposure. However, in cases where exposure has occurred prior to any type of remedial intervention (e.g., through workplace exposure) and the question of "how much?" is being asked, measurements of chemical levels in the body provide the best answer.

Measurement techniques for evaluating chemical levels in the body are available for a number of occupational chemicals and are used to screen workers for occupational exposure to these chemicals. Fewer techniques are available for chemicals that are potential problems at hazardous waste sites (100). One chemical for which environmental risk is evaluated based on body burden (and the only chemical at hazardous waste sites that is evaluated in this manner on a regular basis) is lead (11,63,101). The potential for exposure to lead and the health hazards associated with lead are expressed in terms of allowable levels of lead in human blood in µg lead/deciliter of blood (µg/dl). The EPA (12) has determined that very low levels of lead in children can pose risks of learning impairment and over time has lowered the target level of lead in blood to its current level of 10 µg/dl. Environmental (non-site-related) lead such as lead in water supplies and in food, can contribute a substantial portion of the total body burden (63). Consequently, any lead from a hazardous waste site is adding to an existing baseline body burden.

Because the purpose of risk assessment is to help prevent or limit exposure, measuring blood lead levels after exposure may have occurred is not the ideal approach for all sites. Studies have been conducted that evaluate the correlation between blood lead levels and lead levels in air, water, food, and soil. These studies have been used as the basis for models to predict blood lead levels based on measured lead levels in environmental media (11,63,101). A problem with the models is that they do not necessarily account for differences in bioavailability of lead from different sources, which is the key advantage of using body burden measurements in the first place. However, Chrostowski (102) reviewed blood lead levels collected from children in a western mining town and compared measured blood levels with expected blood lead levels predicted using the EPA's Lead Biokinetic/Uptake Model; the results indicate a good correlation down to soil levels of around 1,000 mg/kg.

The models still retain some advantages over standard exposure assessment approaches in that they allow the contribution to lead body burden from multiple sources to be considered and make better use of empirical information. Because of these factors, the models tend to give results that are somewhat less conservative than results obtained using the standard CDI:RfD approach. For example, allowable soil levels of between 100 and 500 mg/kg for lead were determined using an ADI approach, while a level of 1,000 mg/kg was determined to be acceptable using a correlation model (11).

1.3 Multiple Chemical and Pathway Exposures

If more than one carcinogen is present, risks associated with the individual contaminants are summed to determine the total cancer risk associated with exposure, as suggested in the EPA (61) guidelines for Health Risk Assessment of Chemical Mixtures. For noncarcinogens, the HQs are summed to determine if combined exposure may pose a health concern (3,61). It should be noted that the summed CDI/RfD ratio (termed the hazard index or HI) is used only as an indication of possible hazard, as summation is only valid if there are no synergistic or antagonistic interactions among the summed compounds and if they have the same mechanism and site of action. As noted in Chapter 3, the approach of summing risks and HQs is an oversimplification of a complex problem. However, this approach is probably useful in providing a rough estimate of likely maximum effects.

In addition to summing the risks posed by different chemicals, individuals may be exposed to more than one medium by more than one route. For example, a child may receive both on-site and off-site exposure to soils and may also be exposed to airborne contaminants. As another example, under a residential use scenario, an individual could be exposed to chemicals in water through ingestion, inhalation while showering or using household appliances, and dermal exposure while bathing; in addition, exposure to chemicals in soils via gardening or consuming homegrown vegetables could occur. However, exposure assumptions used in most risk assessments are generally designed to provide conservative estimates of the potential for exposure to chemicals in a particular medium, and it may not be realistic to sum exposures across media. For example, it is unlikely that an individual child would ingest soils from an on-site residential area for 365 days/yr and also ingest sediments 100 times/yr from a nearby stream. A qualitative discussion of the potential impact of the multiple possible exposure routes should be included as part of the uncertainties section.

2.0 REGULATORY CONSIDERATIONS

An upper-bound excess lifetime risk level of a one in one million (10^{-6}) is often used at hazardous waste sites as a risk level goal for carcinogens. Regulatory criteria are generally established within a risk range of between one in ten thousand and one in one million (10^{-4} and 10^{-6}), based on a consideration of practicality, size of the exposed population, etc. This policy is consistent with the National Contingency Plan (NCP) developed by the EPA:

> For known or suspected carcinogens, acceptable exposure levels are generally concentration levels that represent an excess upper bound lifetime cancer risk to an individual of between 10^{-4} and 10^{-6} using information on the relationship between dose and response. The 10^{-6} risk level shall be used as the point of departure for determining remediation goals for alternatives when ARARs are not available or are not sufficiently protective because of the presence of multiple contaminants at a site or multiple pathways of exposure;
>
> (40 CFR 300.430).

An excess lifetime cancer risk level of 10^{-6} means that an individual exposed to a given level of the carcinogenic chemical for a lifetime would have an increased probability of developing cancer (as compared to an unexposed individual) of one chance in one million. The actual target risk level of concern for a site is determined by the risk manager (usually the appropriate regulatory agency). Generally either 10^{-5} or 10^{-6} is used as a target risk level by state and federal agencies and several states such as Washington State incorporate both, with total risk required to be less than 10^{-5} and no single compound exceeding a risk of 10^{-6}. The final risk criteria selected should consider such factors as natural background, practicality, other factors influencing potential exposure, etc.

For noncarcinogens, most states and EPA assume that if the HI and HQ are less than one, then the potential for noncarcinogenic risks is small. However, the EPA Office of Drinking Water (103), in establishing Maximum Contaminant Levels (MCLs; Primary Drinking Water Standards) uses an uncertainty factor of 5 to account for non-water sources of chemicals (e.g., the diet or other environmental sources). At least one state, Massachusetts has also considered other possible sources of chemicals in assessing hazard. Massachusetts assumes that only an HI or HQ less than 0.2 poses an acceptable health hazard.

3.0 PRESENTATION

The ultimate goal of remediation, and consequently, of risk assessment as a tool of remediation, is to solve a problem. This requires that the risk assessment be readily understandable to the people making decisions on the site. The most scientifically valid risk assessment is useless if its results are difficult for risk managers to understand.

In presenting data to other scientists (including regulators), most scientists (e.g., risk assessors) tend to gravitate towards the more is better school of data presentation. While it is true that regulators need a complete picture of a site and adequate supporting data, it is not necessarily true that a complete picture is best provided by a large database. CEOs typically get one page summaries of key developments and base major decisions on these memos. Like these CEOs, hazardous waste site regulators tend to have substantial work loads and significant time constraints, and consequently, need to have data that are complete but summarized in a readily comprehensible format. The regulators are in a difficult position: they often have lots of sites, little time to evaluate these sites, and are unlikely to have expertise in the many disciplines that influence their jobs. In addition, while protecting public health and the environment is a priority, regulators are required to take a pragmatic approach to problem solving. Consequently, regulators tend to get caught in the middle of opposing factions such as between site operators and the public, between opposing public viewpoints, or between opposing property owners.

Regulatory agencies get approval for their actions ultimately from the public, and consequently, anything that can be done to the risk assessment to increase its acceptability is useful. Regulators and the public may deny certain actions based on a lack of understanding of the conservatism in the risk estimate, a lack of confidence in the proposed solution, or because of a lack of trust. Such a response (i.e., your risk assessment-based cleanup goals are judged unacceptable for this site), when the risk assessor has put considerable effort into a risk assessment, is disheartening at best, and it is easy to criticize the regulator for a bad decision. However, in most cases, if in their shoes we might make the same decision. Adequately summarizing and explaining issues in the risk assessment while being mindful of the needs of regulators to respond to public concerns, can be the most important part of a risk characterization.

In presenting risk information to the public, risk perception can be as important as the risk calculations. For example, cyanide in soils at 7,000 mg/kg may be protective of human health from a technical perspective, but cyanide in soils at a level of over 1000 anything (e.g., mg/kg, ppm, ppb) may be perceived as unacceptable from a public relations perspective (104). It is important to consider the reception of the value prior to presentation in order to either address anticipated concerns by altering the presentation and, hopefully, the perception or, if the presentation cannot be altered, to ensure that the presenter is prepared to deal with the issues that are raised.

Two ways to make the risk assessment more understandable to the public and to increase the acceptability of its results are to provide points of comparison and to present alternative descriptions of the results (104). Providing risk-based values developed for other sites or by other agencies can give the public and regulators some degree of confidence that values presented in the risk assessment under review are consistent with other criteria. If the results are not consistent, the reasons for the inconsistency should be clearly described. Making decisions without knowledge of prior action on similar sites is likely to be very difficult, and regulators and the public may not have ready access to such information. Providing this information can facilitate review.

Presenting the results of the risk assessment in more than one way is another approach that can be used to increase the chance of a risk assessment being accepted. Different audiences have different analytical approaches to problems and presenting an answer in more than one manner increases the chance of addressing the reviewers' individual concerns. As an example, presenting the amount of chemical in soils and the amount of material likely to be emitted from a site in terms of total chemical mass may be helpful. Comparing the level of a target chemical for which criteria are not available to available criteria for a similar chemical is another approach that may be useful in some cases. The key factor to remember is that the risk characterization, in addition to providing the answer for the risk assessment, is also the section that needs to solve the problem of getting the appropriate message delivered to its intended audience. The most accurate risk assessment that shows no further action is needed is useless if its intended message is not understood by the public and regulators making decisions on the site.

4.0 PERSPECTIVE

To summarize, the numeric calculation of risks for a hazardous waste site is generally a fairly simple and straightforward process that consists of combining results from the exposure assessment and toxicity evaluation sections of the risk assessment. The function of the risk characterization is not only to present these numeric solutions, but to address the uncertainty inherent in the process and finally to provide this information in an understandable form. Explaining and describing the uncertainty in a risk assessment, and convincing an audience of the appropriate action in the face of this uncertainty, is often a difficult and complicated task.

5.0 GENERAL REFERENCES

Cohrssen, J.J. and Covello,V.T., *Risk Analysis: A Guide to Principles and Methods for Analyzing Health and Environmental Risks.* U.S. Council on Environmental Quality. National Technical Information Service (NTIS) Order No. PB 89-137772 (1989)

Environmental Protection Agency (EPA), Superfund Public Health Evaluation Manual. Office of Emergency and Remedial Response, Washington, D.C. (OSWER Directive 9285.4-1) EPA 540/1-86/060 October (1986)

Environmental Protection Agency (EPA), Risk Assessment Guidance for Superfund: Human Health Evaluation Manual, Part A (Interim Final). Office of Solid Waste and Emergency Response, Washington, D.C. OSWER Directive 9285.701A July (1989)

Paustenbach, D.J., ed., *The Risk Assessment of Environmental Hazards: A Textbook of Case Studies.* John Wiley & Sons, New York. (1989)

6.

UNCERTAINTY IN RISK ASSESSMENT

Uncertainty is inherent in risk assessment (105). Toxicity values designed to protect a heterogenous human population exposed to low levels of chemicals are based on studies that expose homogeneous rodent populations to high levels of the target chemical. In an exposure assessment, the actual amount of material an individual is exposed to depends on many factors which may be difficult or impossible to measure accurately, including the duration and frequency of contact with chemicals, the location of a residence, specific location of chemicals, and age, body weight, sex, and activity patterns of the individual. Given this inherent uncertainty, it is not surprising that regulatory agencies and others responsible for protecting public health, generally use approaches and assumptions that are designed to be health protective, that is, they are designed to produce estimates of exposure that overestimate rather than underestimate exposure and risk.

A problem with the use of upperbound estimates and assumptions throughout a risk assessment, is that large overestimates of exposure and risk can be produced. Such large overestimates are a problem, as economic resources are limited, and overprotecting one aspect of life can lead to under-protecting other, potentially more important concerns (106). This message was brought home on a recent train ride along the Northeast Corridor, where small piles of neat, plastic-covered soil that probably cost their owners hundreds of thousands to remediate, sit in sharp contrast to horribly rundown tenement housing. Comparison of the public health risks of contaminated soil with risks associated with horrible housing suggests an inappropriate allocation of funds. This is not to say that hazardous waste sites should not be remediated until better housing is provided for the poor but rather to note that it is important to make a good effort to refine risk assessments to limit uncertainty and consequently to limit the extent of remediation. The purpose of this chapter is to first describe the problem, i.e., to identify the types of uncertainty in a risk assessment, and then to present approaches for addressing this uncertainty.

1.0 INTRODUCTION

Major sources of uncertainty in most risk assessments include:

- the exposure scenario
- sampling and analysis
- measurement of exposure parameters
- estimation of chemical behavior in the environment
- toxicology (both qualitative and quantitative aspects)

In other words, sources of uncertainty in a risk assessment generally include almost all aspects of the assessment. The sources of uncertainty are discussed in previous chapters of this book. This chapter focuses on characteristics of uncertainty in risk assessment and the approaches used to address this uncertainty.

2.0 QUALITATIVE CONSIDERATIONS

A risk assessment has two main components, the exposure assessment and the toxicity assessment. The actual risk characterization (the third component of the risk assessment) is the combination of information from the exposure assessment and toxicity assessment. The types of uncertainties in these components differ somewhat but generally can be placed in three categories:

1. Uncertainty caused by a lack of data on a specific parameter,
2. Uncertainty associated with normal variability in the parameter, and
3. Uncertainty introduced by site-specific or individual-specific conditions that may or may not be quantifiable.

The first of these factors affects the amount of weight (trust) that should be placed on the particular exposure parameter or toxicity value.

The second factor addresses the likelihood of the assigned value of a toxicity criterion or exposure parameter applying to any exposed individual. An important consideration is that many parameters are dependent; for example, the amount of air breathed by an individual is a function of the size of that individual and the effect of a particular chemical may be strongly linked to aspects of an individual's diet. The third factor affecting uncertainty, site- or individual-specific conditions, influences whether or not the standard default values are applicable to the site or individual. The interaction of these factors influences the overall uncertainty in the estimate.

2.1 Toxicity Assessment

Because of the differences in how carcinogens and noncarcinogens are treated, the first step in a toxicity assessment involves establishing whether or not a chemical is a carcinogen. This categorization is the first source of uncertainty in the toxicity evaluation. The EPA, the International Agency for Research on Cancer (IARC), and the National Toxicology Program (NTP) have developed ranking procedures to describe the weight of evidence for carcinogenicity. The EPA (12) approach, which is the most frequently used in risk assessment for hazardous waste sites, ranks chemicals into six categories (Table 1). This classification scheme has been useful to EPA in organizing their approach to assessing cancer risks. Only limited effort has been directed at ranking the quality of the data for noncarcinogens.

TABLE 1
Weight of Evidence Ranking Scheme for Carcinogenicity

Group A, Human Carcinogens - Chemicals that are considered to be carcinogenic based on sufficient evidence of carcinogenicity from epidemiological studies.

Group B1, Probable Human Carcinogens - Chemicals considered to be carcinogens based on limited evidence from human studies and sufficient evidence of carcinogenicity from animal studies.

Group B2, Probable Human Carcinogens - Chemicals for which there is sufficient evidence of carcinogenicity in animals but no evidence or insufficient evidence from human epidemiological studies.

Group C, Possible Human Carcinogens - Chemicals for which there is only limited evidence of carcinogenicity from animal studies.

Group D, Not Classifiable as to Human Carcinogenicity - Chemicals in this category have inadequate data on carcinogenicity.

Group E, Evidence of Non-Carcinogenicity in Humans - Chemicals that have shown no evidence of carcinogenicity in adequate human or animal studies.

For both carcinogens and noncarcinogens, uncertainty associated with normal variability among animal species in response to a chemical has generally been addressed by the policy of using the most sensitive sex of the most sensitive species as the basis for the criteria. This approach ensures that to the extent possible, the derived toxicity criteria will not be under-protective. Dourson (36) has developed a scheme to attempt to use more of the data in developing toxicity criteria for noncarcinogens but the approach has not been widely accepted.

Another approach to addressing the variability among species in response to chemicals is to attempt to understand differences in the physiology of the species and use this information in the assessment (29,107,108). This type of approach has recently received much attention in cancer risk modelling. Physiologically based pharmacokinetic modelling (PB-PK modelling) in theory should reduce the level of uncertainty in cancer risk modelling (44). However, a substantial amount of uncertainty is still associated with the results from the PB-PK model.

Uncertainty introduced by site-specific or individual-specific conditions that may or may not be quantifiable include conditions that impart a specific toxicity to an individual, such as Wilson's disease, in which individuals with the disease are highly susceptible to copper toxicity. There is increasing evidence that such factors may play an important role in determining who contracts a certain disease or displays a particular response to a chemical agent. Other factors that may influence a response include diet, and the effects of differences in exposure to other chemical agents, including naturally occurring chemicals in foods, the air, and water.

2.2 Exposure Assessment

In toxicology, criteria are derived from studies on homogeneous animal populations and most uncertainty is associated with questions about the applicability of these animal data to humans. In exposure assessment, studies are generally conducted on humans and uncertainty tends to be associated with variability in the human population. Uncertainty can be caused by a lack of data (i.e., too few studies) or by differences in populations (people in Georgia probably drink more liquids, on average than people in Washington State) or by the normal variability inherent in the exposure parameter. Finally, site-specific conditions influence whether or not the exposure assessment may even be applicable at a particular location. The interaction of these types of uncertainty influences the overall uncertainty in the exposure estimate.

Quality of the Data. The amount of data available to serve as the basis for estimating each exposure parameter varies substantially. For such factors as body weight and inhalation rates, large amounts of data are available, while for other parameters such as the amount of soil ingested by adults or the amount of industrial building surface contacted by workers there are almost no data. Standard practice in risk assessments for regulatory agencies is to ensure that risks are not underestimated by using upper-bound estimates of exposure parameters. As risk assessment develops as a science and attempts are made to more accurately define actual risks, it becomes increasingly important to differentiate between exposure parameters that are based on good quality data and those parameters that are simply based on the best guesses of researchers or regulators.

As noted above, ranking of data quality (i.e., the weight of evidence ranking) has been well developed in carcinogenicity assessment. A similar approach is not currently available for ranking exposure parameters but would be useful 1) in evaluating the weight of evidence and consequent uncertainty in exposure values and 2) in focusing research efforts on critical exposure parameters.

Evaluation of the weight of evidence for carcinogenicity is a relatively simple process, primarily because a dichotomous response is being sought. Policy decisions are made as to what constitutes a positive response in epidemiological studies, animal bioassays, and genotoxicity tests. A chemical generally either is or is not positive in a particular test for carcinogenicity (some equivocal results are obtained but criteria are commonly developed to categorize even these as either positive or negative). Once the response in the various species has been evaluated, chemicals can be fairly readily classified using the weight of evidence ranking scheme.

For exposure assessment, the process is substantially more complicated, because the accuracy of a particular value is being evaluated. The ease in determining the accuracy depends on several factors, including: variability in individual exposure, number of studies assessing the parameter, and the scientific merit of these studies. For a factor such as body weight with limited variability (i.e., with variability generally in the percent range rather than in the order of magnitude range), fewer studies are required to ensure that a reasonably accurate estimate of the actual parameter value is obtained. For factors with larger and potentially regional variability such as soil ingestion, more and higher quality studies will be needed. Therefore, variability, study size, and quality need to be included in a ranking scheme.

A preliminary ranking scheme is presented in Table 2. The number of studies required to classify a particular exposure parameter are somewhat arbitrary but give some idea of the magnitude of data needed to reach a decision. Certain exposure parameters, such as those that are difficult to measure and that have fairly low variability, may be reasonably ranked in a higher category if general opinion supports such a ranking.

TABLE 2
Proposed Ranking Scheme for Exposure Parameters

Group A - Either 1) Over 1,000 measurements, one high quality study and at least two supporting studies or 2) Over 100 measurements, one high quality study and two supporting studies and limited variability (Less than factor of 5).

Example: Body weights (good data from many people)

Group B - Over 100 measurements, and either several good quality studies that are supportive or one high quality study and at least some supporting evidence.

Example: Soil ingestion rates for children (4 studies; limited variability)

Group C - Limited empirical evidence available; or substantial variability in the available data.

Example: Fish consumption rates (several studies available but rates differ substantially)

Group D - No empirical evidence supporting value.

Example: Contact rates with exposed surfaces

Variability in the Exposure Parameter Estimate. The actual amount of material an individual is exposed to via any particular exposure route depends on the individual's frequency, extent, and duration of exposure, which in turn depends on many factors (e.g., location of residence, age, body weight, sex, and activity patterns). Patterns of exposure are highly variable among individuals. For example, the amount of soil ingested by a child may vary from almost none for a child playing on a grassy area or indoors in a very clean house, to somewhat less than 100 mg/day for most children, to relatively large amounts (5,000 mg or more) in children who are intentionally eating soil (77,89).

Variability tends to be rather small for parameters such as weight and age and for those parameters that are essential for life such as breathing or drinking liquids. On the other hand, variability can be large for parameters such as soil ingestion or fish consumption and can include no exposure in many cases. Also, as noted above, variability is not necessarily independent. The amount of soil ingested tends to be inversely related to body weight as children tend to ingest larger amounts of soil than adults. On the other hand, water intake, inhalation rate, and food consumption are all directly related to body weight.

Variability in the exposure parameters can be characterized by describing the distribution of the values. Several researchers have presented distributions for exposure parameters as part of Monte Carlo analyses or have presented ranges of potential exposure values (8,98,109,110). Burmaster and colleagues have been engaged in developing distributions for key parameters such as soil ingestion rates, water consumption, body weights, and fish consumption rates (98,109). The exposure parameters, an estimate of the central tendency of the parameter, an estimate of the distribution of the value and an indication of any covariants all need to be noted. In addition, it is important to know if the variable is a primary factor in the exposure assessment and directly influences risk or if it is a secondary factor for which major variability (or inaccuracies) will minimally impact the final exposure results.

Unquantifiable Site-Specific Factors. Site-specific factors can affect the potential for exposure in a dichotomous manner (no exposure vs. some exposure). For example, soil ingestion is a pathway of concern for certain persistent compounds and metals. However, inclusion of this pathway in a risk assessment assumes that site soils are available to be ingested or will be available in the future. If soils are covered by asphalt or grass, this exposure may not be possible. In addition to having a dichotomous effect, site-specific factors may have a graded effect on the potential for exposure. The location of areas of higher and lower contamination in relation to future areas of exposed soil can substantially influence exposure (76). Dealing with the site-specific factors potentially influencing exposure is necessary but may be essentially impossible to do with any real degree of accuracy. This is particularly true in light of the fact that most risk assessments assume use of a property for long periods of time (e.g., 30 years for most residential use scenarios).

3.0 QUANTIFYING UNCERTAINTY

So far this chapter has addressed types of uncertainty in the various sections of a risk assessment. This uncertainty is inherent in the process but it can be minimized to some degree and perhaps more importantly, its effects can be minimized. Engineers have been working for years to minimize uncertainty using techniques such as Monte Carlo analysis (110), and to minimize the effects of the uncertainty using techniques such as the observational method (111, 112), in which data are collected to allow midstream scoping and adjustment. Both of these procedures should serve to ultimately decrease the costs of remediation through decreasing the effects of uncertainty.

As noted in the opening section of this chapter, the potential costs to public health of overspending money on remediation activities because of an overly conservative risk assessment are substantial. In the private sector, jobs and health insurance may be unnecessarily lost or diminished. In the public sector, public health projects such as pre- and post-natal care for the poor may need to be curtailed for lack of funds. More accurately defining health risks associated with hazardous waste sites can lead to a better balancing of costs and benefits to society as a whole.

3.1 Semi-Quantitative Evaluation

From a practical standpoint, it is often better to have a clear knowledge of the direction of uncertainty than an actual measure of the uncertainty. For example, when writing a check it is more comforting to know for certain that you have enough money to cover the cost (or that some uncertainty factor such as overdraft protection has been built in) than to be 95% sure that you have enough. If money is not a major constraint, then it is always better to be certain of a positive outcome. However, because resources are generally limited (e.g., money left in a checking account is not available to pay other bills or to go into a higher-yielding money market account) actual measures of uncertainty become more important as part of a cost/benefit assessment.

For hazardous waste site risk assessments, uncertainty is only a major issue if a worst case-type analysis suggests that risk may be present. If a totally unrealistic but clearly conservative estimate (i.e., clearly an overestimate) of risk shows that no problem exists, then refining the estimate is unimportant. Consequently, preparing a worst case-type analysis of a problem may be a useful first step, particularly as conservative screening-type assessments generally are much less costly than detailed evaluations. However, one of the golden rules (Murphy's Law) of site work is that all risk estimates tend towards a number just slightly above target risk goals. The screening risk assessment can still be useful in focussing evaluation on critical pathways, chemicals, etc.

An important point to keep in mind is that even upper-bound risk estimates developed using standard procedures do not necessarily protect all individuals. In some cases, individual sensitivity to chemicals will be orders of magnitude different from that of the general population and it will be essentially impossible to ensure protection of all individuals without spending clearly inordinate amounts of money or removing valuable material from use. In other cases, specific situations will lead to exposures outside the "normal" range but may need to be considered on a site-specific basis. For example, exposure assessments do not typically address habitual pica in which a very small subgroup of children ingest up to several grams of soil or other materials for a short period of time. This is unlikely to be a problem if the chemical of concern is only chronically toxic. However it may be useful to attempt to include these routes if working in a lower socioeconomic area (children in these areas are more likely to exhibit habitual pica) or in an area where the ingestion of soils is culturally encouraged. As another example, in certain areas, subsistence fishing is important and use of the EPA maximum fish intake rate may not be sufficiently protective.

In attempting to provide a qualitative to semi-quantitative evaluation of the uncertainty in the risk assessment, the specific factors influencing uncertainty are listed and their effects on the assessment are noted. For example, Tables 3 and 4 illustrate two approaches to detailing uncertainty to at least a limited degree. Another approach that has been used is to prepare both average and maximum likely estimates of risk and compare the differences. As an example, target groundwater concentrations for a carcinogen can be calculated using the following standard equation:

Eq (15)

$$C = \frac{TR \times BW \times AT \times 365 \text{ days/yr}}{EF \times ED \times SF \times IR \times FI}$$

Where:

C = Target water concentration (mg/L)
TR = Target cancer risk (10^{-6})
BW = Body weight (70 kg)
AT = Averaging time (70 years)
EF = Frequency of site contact (350 days/365 days)
ED = Exposure duration (30 years)
SF = Slope factor ($(\text{mg/kg/day})^{-1}$)
IR = Water consumption rate (2 L/day)
FI = Relative fraction absorbed from water (100%)

TABLE 3
Example of an Uncertainty Table for Risk Assessment[a]

	EFFECT OF EXPOSURE		
ASSUMPTION	Potential Magnitude for Over-Estimation of Risk	Potential Magnitude for Under-Estimation of Risk	Potential Magnitude for Over- or Under-Estimation of Risk
Water is contaminated at a constant level for the entire period of exposure	Moderate		
Measured levels in water are accurate			Low
Modelled exposure point concentrations are accurate	Moderate		
2 Liters of water consumed per day			Low
All water consumed comes from the site	Low		
Individual will reside in the same house for 30 years			Low
Body weight is 70 kgs			Low
Lifetime is 70 years			Low

[a]Adopted from EPA (1989)

[b]"As a general guideline, assumptions marked as "low", may affect estimates by less than one order of magnitude; assumptions marked "moderate", may affect estimates by between one and two orders of magnitude; and assumptions marked "high", may affect estimates by more than two orders of magnitude" (EPA 1989).

TABLE 4
Uncertainties in the Risk Assessment

Uncertainty Factor	Effects of Uncertainty on Risk	Comments
Water is contaminated at a constant level for the entire period of exposure	May overestimate risks	Effect may be quite substantial particularly for degradable chemicals
Measured levels in water are accurate	May over- or underestimate risk	Analytical error often on the order of +/-30%; sampling error may also influence outcome
Modelled exposure point concentrations are accurate	May overestimate risk	Depending on level of detail in the model, effect could be substantial
2 Liters of water consumed per day	May over- or underestimate risk	Variable depending on climate
All water consumed comes from the site	May overestimate risk	Effect could be substantial for people who work outside the home
Individual will reside in the same house for 30 years	May over- or underestimate risks	Probably overestimates risk for people; effect probably small
Body weight is 70 kgs	May over- or underestimate risk	Effect probably small; most adults probably weigh between 100 and 200 pounds (+/-50%)
Lifetime is 70 years	May over- or underestimate risk	

The values presented above are the standard exposure assumptions used in EPA risk assessments. The difference between the outcome of a risk assessment using this approach and the best estimate value (central tendency number in EPA's new guidance) developed using best estimates for each parameter gives some idea of the uncertainty in the assessment.

Parameter	Standard	Best Est.	Difference
Water consumption rate	2 L/day	0.2 L/day	10X
Fraction absorbed from water	100%	50%	2X
Frequency of site contact	350 days/yr	100 days/yr	3.5X
Exposure duration	30 years	7years	4X
Body weight	70 kg	70 kg	-
Averaging time	70 years	70 years	-
Total Difference:			**280X**

Based on this analysis, a best estimate differs from the EPA estimate by approximately a factor of 300. However, as illustrated by this approach, it is not possible to determine whether the EPA estimate is the 95% upper confidence limit (UCL), the upper 99% UCL or the upper 99.99% UCL. Without this type of information, risk managers are only able to guess at the actual degree of protection being afforded by the risk assessment.

3.2 Probabilistic Risk Analysis

Recently, risk assessors have begun using Monte Carlo estimation techniques to attempt to more accurately define the uncertainty in risk assessments (98,110) and to allow risk managers to more accurately characterize risk. Approaches to more accurately quantifying risks such as the use of Monte Carlo analysis can ensure that health and the environment are protected, while decreasing the need to substantially overestimate risks. These more detailed analyses of uncertainty were initially not widely accepted by regulators, probably for some of the same reasons that standard risk assessments may not be accepted (a lack of time, expertise, or resources to adequately review the work). However, knowledge of risk assessment procedures is expanding and more and more EPA Regions are considering the use of more advanced procedures such as Monte Carlo analysis.

As noted previously, combining upper-bound estimates yields an overall overestimate of likely risks that is the product of the individual uncertainties. Monte Carlo analysis is a process whereby distributions of the various parameters that make up the risk assessment are combined instead of point estimates of these parameters. For example, instead of a single point estimate of soil ingestion such as 200 mg/day, a distribution based on a median value of 60 mg/day and a standard deviation of 125 mg/day would be used to estimate exposure (109). The distributions are combined to produce a final probability distribution of risks. From this probability distribution, the risk level associated with a specified upper confidence limit (e.g., 95%, 99%) can be determined. For example, a risk manager may determine that it is important to ensure that the risk estimate be protective for 99% of the population and the risk associated with this value can be determined.

Monte Carlo analysis is a powerful tool in the right circumstances but it is subject to some limitations. These include an unwillingness on the part of regulators to accept its use (many regulators are just getting to the point of considering the use of risk assessment in decision making and the use of a fancy statistical approach may be difficult to swallow). Another limitation is the appearance of precision that is afforded by the procedure (Monte Carlo results are somewhat more accurate but look a lot more accurate than standard values). Finally, the procedure cannot account for site-specific factors such as current and future cover at a site. However, keeping these limitations in mind, the use of Monte Carlo-type analyses can greatly improve the accuracy of risk assessments and more importantly, can decrease the large degree of overestimation required to account for the uncertainty inherent in the risk assessment process.

3.3 The Observational Method

In general, activities at a hazardous waste site proceed in the general order of investigation (possibly with several phases), risk assessment (including selection of appropriate cleanup criteria), evaluation of remedial alternatives, design of the selected alternative, and implementation of the remediation. While this approach is appropriate in some cases, in many cases, a more interactive approach involving communication among people working on the separate parts of the process would greatly improve the final product in terms of cost and human health protection.

The Observational Method is an engineering approach suggesting that design of a system should be somewhat flexible to allow changes based on observations from the field (111,112). The approach assumes that uncertainty is inherent and suggests that rather than over-engineering to ensure safety, in some cases it may be possible to use midpoint adjustments to ensure final results are achieved. As an example, consider a site where the primary goal is to ensure protection of an off-site groundwater well. Simple modelling suggests that there is a chance that if no action is taken the well could be contaminated. Two options would typically be employed in this situation. One is to conduct an elaborate modelling effort to show definitively (hopefully) that the well would not be impacted. The other is to design a groundwater control system to prevent the contaminant from migrating to the well. The latter of these options is clearly effective but costly. The former is probably (again, hopefully) less costly but may not be acceptable to all parties. The observational method offers a third approach. In this case , a monitor well could be installed midway between the known contamination and the well and this well monitored on a time schedule based on the results of the simple model. Results from this well will allow an accurate determination of what actually happens to groundwater at the site. In some cases, this accurate information may still require implementation of a groundwater control system. However, in many cases, the data will show that attenuation (degradation and binding to soils) of the chemical in groundwater is greater than expected and no further action will be necessary.

The Observational Method can be useful in risk assessment in several ways. First, as noted above, if a risk assessment is designed to allow revision based on the results of future groundwater monitoring rather than on modelling, fewer assumptions are necessary and risks can be depicted more accurately. Second, maximum exposure scenarios are unnecessary, if it is understood that before a site is used, the risks posed by that use must be assessed. Finally, many on-site remedial techniques do not achieve total destruction of chemicals and it may be difficult or impossible to estimate final levels that will be achieved at a site. However, considering that the remedial action is generally much more intrusive than natural conditions, it is likely that if residual levels of the chemical cannot be removed by active treatment, these residuals will pose an insignificant risk and can be left in place at whatever level is achieved.

As an example of the use of the Observational Method in setting final remediation goals, consider the following case. Trichloroethylene (TCE) initially present at 100 mg/kg, may only be treatable *in situ* to 5 mg/kg in a tight, organic-rich sandy clay soil, while cleanup levels of 0.5 mg/kg are achievable in sandy soils. If it takes 60 days to achieve the 5 mg/kg cleanup in organic clay and only 10 days to achieve the 0.5 mg/kg in sand, the residual TCE in the sand may be more available (and more of a threat to neighboring groundwater) than the higher level that is more tightly bound. Non-site-specific cleanup levels would not account for this tight binding, and even levels developed based on the results of a laboratory treatability study might not indicate the full binding ability *in situ*. However, establishing conditional cleanup goals, and only establishing final remediation levels at the end of the project would allow the full effect of the binding to be recognized and would consequently decrease the uncertainty in the site remediation. This approach clearly requires a degree of flexibility on the part of the regulators involved with the project, who will have to be willing to accept such things as not knowing final cleanup goals until close to the end of the project (or at least until several areas of the site have been treated) and different cleanup goals for different sites. For the parties initiating the cleanup, there will be some uncertainty as to the acceptability of an unknown remediation goal to the regulators. However, the alternative, over-engineering the remediation (e.g., excavation to a predetermined remediation goal and off-site incineration) may pose inordinate costs.

4.0 PERSPECTIVE

Obviously, dealing with uncertainty is an inherent part of the risk assessment process. Addressing the uncertainty requires a balanced approach that considers not only the information available to address the uncertainty but also the effect uncertainty in a particular parameter has on the results. In some cases, substantial effort has been focussed on defining values that have only a minimal impact on the overall results of the assessment. In other instances, little or no data are available for parameters that greatly affect the risk estimate. Characterizing the weight of evidence supporting specific exposure and toxicity parameters may help to better identify data gaps and facilitate the process of decreasing uncertainty in individual values and ultimately in risk estimates.

Addressing uncertainty in risk assessment also requires attention to accurately defining the factors affecting uncertainty for the particular site. Some estimate of the magnitude of the impact of the uncertainty is also useful, be it a semi-quantitative evaluation or a more detailed probabilistic assessment (Monte Carlo analysis). In addition, use of more detailed site data to accurately define site parameters and perhaps to allow mid-stream adjustments to the risk assessment or to remedial actions may substantially reduce uncertainty and thereby improve cost-effectiveness.

As a final note, perhaps the main cause of the overemphasis on minor changes in exposure and toxicity parameters is the often inflexible use of regulatory criteria. Most criteria are treated as levels not to be exceeded, without regard for the huge degree of uncertainty that went into deriving these values in the first place. As an example, an EPA fact sheet for a Superfund site notes that "no treated soil may be returned to the excavation if it has a maximum concentration of PCBs above 49 ppm. (PCB concentrations at 49 ppm or below do not present a risk to human health.)" Thus, according to this information, 51 ppm PCB may present a risk but 49 ppm does not. Given the uncertainty in analytical data alone, this type of statement makes little sense. While limits must be established, it is important to maintain flexibility on a case-by-case basis. For example, in the above situation, it might have been more reasonable to note that EPA classifies soils containing PCBs at levels above 50 ppm as hazardous waste and that levels on site will be maintained at or below this level. This statement avoids the appearance of precision (e.g., 47 or 53 ppm round to be equal to 50 ppm) and does not require the conclusion that a clear distinction can be made as to PCB concentrations that pose a risk versus those that don't. The need for flexibility in regulatory guidance is recognized by several states but a better understanding of uncertainty should lead to increasing the flexibility in criteria.

5.0 GENERAL REFERENCES

D'Appolonia, E., Monitored Decisions. *J. Geotechnical Engineering* 116: 4-34. (1990)

Environmental Protection Agency (EPA), Risk Assessment Guidance for Superfund: Human Health Evaluation Manual, Part A, Interim Final. Office of Solid Waste and Emergency Response, Washington, DC, OSWER Directive 9285.701A, July (1989)

Finkel, A.M., Is Risk Assessment Really Too Conservative?: Revising the Revisionists, *Columbia Journal of Environmental Law* 14:427-467 (1989)

Thompson, K.M., Burmaster, D.E., and Crouch, E.A.C., Monte Carlo Techniques for Quantitative Uncertainty Analysis in Public Health Risk Assessments. *Risk Analysis* 12: 99-104 (1992)

7.

RISK COMMUNICATION

Risk assessment is not rocket science. The best answer is generally the simplest possible answer -- not the response generated using the most complicated approach.

1.0 INTRODUCTION

As noted at the beginning of this book, the primary purpose of remedial action at a hazardous waste site is to protect human health and the environment. However, actions may also be taken to respond to two other requirements, namely, the need to achieve regulatory approval and to address public concerns. Scientists tend to deal in objective truths and therefore are on firm footing in solving problems such as determining if a site poses a risk (even if uncertainty is present). However, we tend to be less comfortable in addressing the subjective concerns associated with such things as public or regulatory personnel's perception of the problem. However, these concerns must be addressed if the site is truly to be remediated. This is not to say we will ever address all concerns or satisfy every individual involved at a site. Remember, if we elect a President with 60% of the vote (i.e., with 40% disagreement) we consider it a landslide. However, a concerted effort must be made to convince as many people involved in the process as possible that the outcome is acceptable.

Risks are a normal part of everyday life and almost everyone makes risk decisions on a daily basis. Decisions about which risks to accept and which to avoid are made based on both our knowledge about the risk as well as on our perception of the risk. People who do not deal with exposure, toxicity, and risk issues on a frequent basis, tend to have a relatively limited understanding of these issues, in much the same way that a toxicologist may be unfamiliar with tax questions. And in the same way that a toxicologist is likely to be very concerned about tax questions (particularly if they come from the IRS), a person with limited risk background is likely to be uncomfortable with risk issues (particularly if they are close to home, as in a neighborhood hazardous-waste site). In each case, the perception of the risk (tax penalties or health risk) is based on our experience and on knowledge obtained from outside sources. The outside sources of information, including newspapers, television, anecdotal information, etc., tend to sensationalize the issues. Consequently, we tend not to hear about the numerous trips to the IRS that involve no penalties or the chemical exposures that do not cause adverse health effects. Rather we are given a somewhat biased understanding of the issues. A clear, unbiased understanding of an issue (be it risk assessment or taxes) is critical to addressing the problem in the most appropriate manner. Consequently, communicating risks and discussing risk issues is at least as important as being able to perform the actual risk analysis.

As noted above, scientists deal in objective truths but issues of perception are often highly subjective (113). Consequently the scientist may present the correct answer to a given problem, without understanding that the message perceived by the audience is somewhat (or substantially) different. As an example, informing an audience at a town meeting that there is no need to be concerned about health risks because soils in their yards only contained dioxins at a level that poses a 10^{-6} cancer risk is likely to result in a different opinion on the part of the audience. The presenter has determined (and probably accurately based on current policy) that this level of dioxin presents a *de minimis* or insignificant risk. And if the audience had been a roomful of like-thinking researchers, the response would have probably involved a substantial amount of head nodding in agreement. However, the audience noted above has a somewhat different knowledge base from which to form their risk decisions. The message that is perceived by this audience is that dioxin, the deadly toxic compound and a carcinogen is present in their backyards. The answer to the question "Does the presence of this chemical pose a risk?" is therefore substantially different from the answer suggested by the speaker. In addition, the presentation of the message, i.e., "there is no need for you to be concerned," is patronizing, and inaccurate in light of the background, knowledge base, and overall concerns of the people receiving the message (114). Acceptable versus unacceptable risks (or risks people are concerned about) include factors such as fears posed by uncertainties in the science (an industrial consultant, the EPA, and a Green Peace representative may all present different "scientifically correct" answers to the same question), concerns about lower property values, etc. Scientists, in attempting to communicate risk issues, need to constantly remember that the subjective perception of an issue can be quite different than the objective answer.

2.0 GENERAL RULES OF RISK COMMUNICATION

The primary rules of risk communication are 1) present information clearly and in a form understandable to the audience, and 2) listen to the audience to make sure that 1) is being realized (114,115). A key component of making information understandable is to remember that people have different backgrounds and perspectives and analyze information differently. Therefore, a clear, precise presentation for one audience may seem relatively illogical to another. Encouraging audience participation through means such as blackboard brainstorming sessions and dividing the audience into groups to discuss issues among themselves can help the risk assessor gain insight into the perspectives and understanding of the audience. Informal audience participation also facilitates interaction by making the risk assessor a resource for information rather than an outsider and adversary.

Covello and Allen (116) noted that there are seven "Cardinal Rules of Risk Communication." These rules are:

1. Accept and involve the public as a legitimate partner,
2. Plan carefully and evaluate your efforts,
3. Listen to the public's specific concerns,
4. Be honest, frank, and open,
5. Coordinate and collaborate with other credible sources,
6. Meet the needs of the media, and
7. Speak clearly and with compassion.

Conversely, researchers at the Environmental Communication Research Program at Rutgers listed "10 Ways to Lose Trust and Credibility (117)". These rules provide a different perspective on what needs to be done to communicate effectively. The rules are:

1. Don't involve people in decisions that directly affect their lives. Then, act defensive if your policies are challenged.

2. Hold onto information until people are screaming for it. While they are waiting, don't tell when they will get it. Just say, "These things take time," or "It's going through quality assurance".

3. Ignore peoples' feelings. Better yet, say they are irrelevant. It helps to add that you can't understand why they are overreacting to such a small risk.

4. Don't follow up. Place returning phone calls to citizens at the bottom of your "to do" list. Delay sending out the information you promised people at the public meeting.

5. If you make a mistake, deny it. Never admit you were wrong.

6. If you don't know the answer, fake it. Never say "I don't know".

7. Don't speak English. When explaining technical information, use professional jargon. Or, simplify so completely that you leave out important information. Better yet, throw up your hands and say "You people could not possibly understand this stuff".

8. Present yourself like a bureaucrat. Wear a three-piece suit to a town meeting at the local grange, and sit up on stage with seven of your colleagues who are dressed similarly.

9. Delay talking to other agencies involved--or other people involved within your agency--so the message the public gets can be as confusing as possible.

10. If one of your scientists has trouble relating to people, hates to do

it , and has begged not to, send him or her out anyway. It's good experience.

Most people who have had to deliver a technical message to a non-technical audience have fallen into one or more of these traps. It is very difficult to communicate technical information to people with widely varying backgrounds, education levels, and interests. Probably the most important factor in getting a message across, is having sufficient time to educate people so that they can make informed decisions based on a reasonable knowledge base. However, sufficient time to allow people to review information and to respond to their questions is often one of the factors in shortest supply. However, because education is probably the critical factor in explaining risk issues, if at all possible, the risk assessor should work to provide educational material to people who may be involved in or affected by decisions at hazardous waste sites.

One important step that may be useful in attempting to communicate risk issues is to present the argument in more than one way (104). This process is illustrated above with the use of the rules and "anti-rules" of risk communication. Some people can clearly understand the steps that need to be followed for effective risk communication by reading the seven rules. However, others (the author included) get a better understanding of what needs to be done by being shown what needs to be avoided. For an example more specific for hazardous waste risk assessment, consider a site that poses an insignificant health risk. Simply presenting the chemicals and associated risk levels may not be sufficient to get a "No Further Action" determination from the regulatory agency or the public. In addition to discussing the risk levels, it may be useful to provide information on levels of contamination at the site (e.g., toluene was detected in 12 of 16 samples at concentrations that averaged 3 mg/kg (ppm) and the maximum value was 12 mg/kg), the total amount of chemical present (e.g., based on mass calculations, only one pint of toluene is present), and to note that the levels present are lower than available standards (state-specific or proposed by other states or EPA) for the chemical (or for comparable chemicals). This additional information is generally simple to collect and allows an audience to view the suggested solution to the problem being addressed by the risk assessment in more than one way.

A final note on presentation format is that the framing of a problem and of the solution to that problem can have everything to do with how the issue is perceived. For example, Slovic (116) notes a study by McNeil et al. in which subjects were asked to imagine that they had lung cancer and were then asked to consider two treatment options: radiation or surgery. The cumulative probabilities were framed in terms of dying or in terms of survival. Some were told that 68% would survive for one year following surgery while others were told that 32% will have died. Framing the problem in terms of dying substantially decreased selection of surgery as a treatment option. Based on this example, it might be better to present risk estimates in terms of survival. In other words, a 10^{-6} cancer risk should be expressed as a potential of not dying from cancer as a result of exposure of 99.9999%. Or assuming that the probability of dying of cancer in the general population is roughly 20% or 0.2, the risk could be expressed as increasing from 0.200000 to 0.200001 or again as a chance of not dying of cancer changing from 0.8 to 0.799999.

Other examples of the influence of framing on decision making are provided by Tversky and Kahneman (118). That framing is a key component of risk communication is important to realize for two reasons. First, the risk assessor should try to frame issues in a way that allows the intended message to be received. Second, it may be useful to show an audience examples of the influence of presentation format on decision making to ensure that all parties are aware of the influence that presentation format can have on their choices.

A knowledge that decisions can be severely influenced by presentation can help dialogue by ensuring members of the audience that their inability to grasp a risk concept may be related to a presentation problem and not to a lack of knowledge. This in turn can help the presenter provide information in a form useable by the audience. Admitting up front that a response can be manipulated may also encourage trust and ultimately acceptance of the intended message as it is evaluated for holes by the audience.

3.0 RISK COMPARISONS

A frequently used approach to addressing risk concerns is to compare the risks at a site with those associated with other common sources of risk, for example, smoking. The problem with this type of comparison is that although objectively similar, i.e., both the site and smoking pose a risk of developing cancer, they are subjectively quite different. The risks differ in being involuntarily versus voluntarily imposed, being unfamiliar versus familiar, having no benefit versus some benefit. Covello et al (119) lists issues that need to be considered in making risk comparisons. These are presented in Table 1.

Risk comparisons may at first appear to be the ideal way around the problem of communicating unfamiliar information to people. However, as noted above there are numerous pitfalls with this approach. One of the best discussions of the issues involved in making risk communications is provided in a Booklet entitled "Risk Communication, Risk Statistics, and Risk Comparisons: A Manual for Plant Managers," prepared by Covello, Sandman, and Slovic for the Chemical Manufacturers Association (119). In this document, the authors describe five ranks of risk comparisons. The authors note that the level of acceptability of risk comparisons by the public varies considerably. Comparisons that are generally acceptable include:

a. Comparisons of the same risk at two different times, for example, a comparison of risk before and after the addition of a treatment system.

b. Comparisons with a standard, (e.g., comparing risks to State or Federal standards), and

c. Comparisons with different estimates of the same risk such as a comparison of the risk estimates prepared by different interested parties.

Comparisons that are generally unacceptable include some of the classic comparisons that are commonly used and that on the surface seem reasonable. For example, the risks posed by chemical x are far less than the risks associated with smoking one pack of cigarettes. The problem with these type of comparisons is that they compare risks that are dissimilar as to voluntariness, familiarity, and personal controllability, and may also simply alienate a subset of the population (smokers in this case) who may feel that they are being patronized or judged. Covello (120) listed ten guidelines that can be used to improve the effectiveness of risk comparisons. These guidelines are presented in Table 2.

TABLE 1

Factors Important in Risk Perception and Risk Evaluation

Factor	Comment
Catastrophic Potential	People have an increased fear of catastrophic events (plane crashes are feared more than car wrecks)
Familiarity	The expression "familiarity breeds contempt" clearly applies to risks (chemical plant workers have higher chemical exposure but lower concern than people exposed to lower levels outside a plant)
Understanding	Things that are not understood cause greater concern than items that are known
Personal Control	People are more willing to accept risks over which they have some control
Voluntariness	Risks that are chosen are more acceptable than risks that are imposed
Effects on Children	Risks that particularly affect children are of primary public concern
Dread	Effects or chemicals with a particular stigma may be perceived as posing greater risk than other, less well known illnesses or compounds
Media Attention	The media can significantly influence the level of concern about a particular issue
Origin	Risks that are viewed as natural or as acts of God are more acceptable than man-made risks

Source: Adapted from (119)

4.0 COMMUNICATING RISKS TO REGULATORS

In addition to the problems of communicating risk issues to local citizens, risk assessors commonly face the problem of explaining the results of a risk assessment to the regulatory managers responsible for making decisions based on the results of the assessment. These regulatory managers are often under severe time and budgetary constraints and may not have the resources in terms of time or expertise to fully understand a risk assessment, particularly a very long, non-summarized document. In addition, as noted in Chapter 5, regulatory agencies are ultimately responsible to the public and they need to be sure that decisions they make will be in the best interest of and acceptable to public. Keeping the needs of the regulator in mind when preparing and presenting the risk assessment report is critical to its acceptance. Adequately summarizing and explaining the risk values, uncertainties and subjective factors influencing risk can be a key element of effective risk communication to regulatory agencies. Again, presenting standards and criteria for the same or comparable chemicals prepared by other agencies may be quite useful. As a warning, although it is sometimes tempting to only present values that support a particular argument (e.g., to only present cleanup criteria that are higher than or equal to the desired cleanup level), to do so is scientifically unacceptable. In addition, it has a tendency to backfire with unpleasant consequences.

TABLE 2

Guidelines for Improving the Effectiveness of Risk Comparisons

1. Target the comparison to the specific audience's needs, concerns, preferences, and level of knowledge.
2. Be specific about the intent of the comparison and caution against unwarranted conclusions.
3. Explicitly state and explain assumptions and uncertainties in the calculations of risks.
4. Discuss the range of possible risk estimates.
5. Avoid comparisons that ignore factors influencing public perceptions of risk and acceptibility, such as voluntariness, fairness, alternatives, and control (Noted in Table 1).

6.	If comparisons that ignore these factors must be used, specifically note the problems with these comparisons.
7.	Focus the comparisons on classes of substances, products, processes, or activities that serve the same function and whose benefits tend to be similar.
8.	Formulate the comparison to illuminate relevant health and environmental consequences, including short-term and long-term effects.
9.	Provide information on the socioeconomic consequences of decisions implied by the comparisons.
10.	Test and evaluate the usefulness of the risk comparisons for the target audience.

Source: (120).

5.0 PERSPECTIVE

Presenting risk issues to an audience that has a limited and probably biased knowledge of risk assessment is a difficult task. This chapter attempts to describe some of the factors that can influence the perception of a message and hopefully provides some guidance on avoiding risk communication pitfalls. However, to reuse an earlier example, remember that in almost any election, more than one candidate (often many more) will receive votes. People have different concerns and needs and even the clearest message is unlikely to be acceptable to all members of an audience. Putting this message more directly, even the best presentation may go awry if members of the audience decide that the message is not one they wish to hear or if they have a different agenda. Risk communication has numerous pitfalls and an intended message will not always be delivered. In order to achieve the highest success rate, the audience has to be carefully considered and allowed to participate in the communication and decision-making process.

6.0 GENERAL REFERENCES

Covello, V.T., Sandman, P.M. and Slovic, P., *Risk Communication, Risk Statistics, and Risk Comparisons: A Manual for Plant Managers,* the Chemical Manufacturers Association, Washington, D.C. (1988)

Slovic, P., Informing and Educating the Public About Risk, *Risk Analysis* 6(4):403-415 (1986)

8.

RISK ASSESSMENT AS A REMEDIATION TOOL

As noted in Chapter 1, the primary purpose of hazardous waste site remediation is to render a site safe for human or ecological populations. Risk assessment is the tool used to measure how clean is clean enough, and consequently, is critical to the remediation process. Hazardous waste site remediation is generally very costly, and more accurate (less conservative) but still, health protective risk-based cleanup goals can substantially decrease costs.

1.0 INTRODUCTION

Because of the uncertainty inherent in risk assessment, it is often possible to provide arguments that have equal scientific merit, for remedial alternatives that range from no action through remediation to former background levels. Associated costs for equally scientifically plausible risk-based goals can range from thousands to millions of dollars. Because the potential future use and actual site characteristics can have such a large impact on the results of the risk assessment, considering risk issues as a factor in final-site use in an iterative manner can decrease uncertainty and substantially influence costs.

The purpose of this chapter is to describe uses of risk assessment in site remediation and to show approaches that can be used to increase the usefulness of risk assessment as a remediation tool. A secondary function of the chapter is to describe the mechanics for calculating risk-based cleanup levels. The final section of the chapter illustrates the use of risk assessment to produce more cost-effective remedial solutions.

1.1 Uses of Risk Assessment in Remediation

Prior to the inception of the laws and regulations controlling the disposal of hazardous wastes, most material that was considered to be toxic was deposited in landfills, burned in incinerators or industrial boilers, or simply poured on the ground. Since the development of rules controlling hazardous waste, the latter of these disposal practices has by and large ceased, but landfilling and incineration continue to be major waste disposal and treatment methods. Both landfilling and incineration involve excavation of contaminated materials, and risk assessment is generally used only in determining how much material needs to be removed, i.e., in establishing how clean is clean enough. Even in these cases, risk assessment has often been used minimally. Site-specific factors are often not considered and the site may be cleaned up to meet standards or guidance that are inappropriate for the particular site. For example, residential use of a property may be unlikely, yet excavation may be based on published guidance levels that assume residential use. Increased use of risk assessment and true, site-specific risk-based cleanup levels should lead to more appropriate and efficacious remediation.

Increasing emphasis is currently being placed on the on-site treatment of hazardous wastes. This emphasis is necessitated by the decreasing availability of landfill space, rules severely restricting the use of landfills, and the high cost of incineration. On-site treatment technologies being used include biodegradation, soil vapor extraction, and stabilization. These technologies, older technologies such as groundwater pumping and treatment, and other new, less well developed on-site treatment technologies, tend to require that low levels of residual contamination be left at the site. For sites employing these non-fully destructive technologies, risk assessment is critical to ensuring that the residual material is unlikely to pose risks to public health or the environment.

1.2 Developing Remediation-Specific Risk Assessments

A risk assessment specifically designed to focus on the appropriate remedial action/cleanup goals for a site does not need to focus strictly on hypothetical maximum use scenarios (as is typically done in a baseline risk assessment) but can be tailored to likely possible uses. For this reason, risk assessment can be used as one element of an iterative process of evaluating possible site uses and remedial options. Factors such as desired future use, available remedial options, and site characteristics will all influence how much risk is present and acceptable, what costs are posed and acceptable, etc.

For example, a very valuable piece of property, or property in an area where high contact is likely (a school yard, for example) will generally require remediation to meet stringent criteria at high overall costs. On the other hand, less stringent cleanup at substantially lower costs may provide the same degree of health protection in a less frequently used area. Looking at the issue in another way, cleanup to X mg/kg will require incineration at a high economic and social cost while cleanup to 10X can be done with biodegradation, and with capping and deed restrictions is likely to pose similar actual risks. The most appropriate option can be selected based on a consideration of desired use, technical feasibility, costs, and risk.

2.0 INCREASING THE USE OF RISK ASSESSMENT IN REMEDIATION

Risk assessments are supposed to answer the question: Does a site pose a risk to public health, welfare, or the environment? In actuality, risk assessments for hazardous waste sites generally address the somewhat different question: Does the site pose a health risk to individuals who might be exposed under a maximum exposure scenario using conservative estimates of exposure? The difference in these questions reflects the uncertainty in risk assessment and a desire on the part of regulators to ensure that public health is protected. Because of this desire to ensure public safety, risk assessment for hazardous waste sites is based in large part on the policy considerations of risk management. This requirement on the part of regulatory agencies that health be protected must be factored in to preparing risk-based remediation goals.

Given that risk management will be a critical component of any hazardous waste site cleanup, perhaps a better question for risk assessments to address is: What is the least amount of chemical in-site environmental media that would possibly harm an exposed individual?

The answer to this type of question at least bounds the risk estimate, in that it is unlikely that significant risks would be posed by levels lower than the target concentration, although higher levels might also not pose significant health risks and in fact for many individuals there may be no actual exposure or risk. Once this question has been answered, a risk management decision can be made as to the appropriate action for the site. If on-site concentrations are all below risk-based concentrations, no further action is probably warranted. If on-site concentrations are several orders of magnitude above the risk-based goals, some remediation is probably necessary. For most sites, some combination of a more refined risk assessment (involving decreasing uncertainty through the use of more site-specific factors and possibly Monte Carlo-type analysis) and remediation will probably provide the best solution to site cleanup. Factors such as desired future use, available remedial options, and site characteristics will all influence how much risk is acceptable, what costs are acceptable, etc. Figure 1 outlines a potential risk assessment decision framework.

The approach of developing cleanup goals based on maximum exposure scenarios and allowing the revision of these values to account for better site-specific information (both better information on potential site uses and exposure, and better measured values for key parameters) is consistent with approaches being used or considered by several states and the EPA (9,121,122,123). These approaches generally present target criteria and a standard approach for developing these criteria, and then also allow site-specific factors to be considered in establishing less stringent, site-specific goals. The approach presented here differs slightly from the previous approaches in removing the need for the standard baseline risk assessment and in focusing the risk assessment on its use in remediation rather than as a stand alone report.

Figure 1

Decision Tree Approach to Site[a] Remediation

1. Do on-site concentrations exceed criteria based on residential use (or other maximum use) scenario? For sites with less than five chemicals of potential concern, use the criteria as developed in the PRG guidance. For sites with over five chemicals, use the PRG values divided by 10 as the target criteria.
Yes -> 2.
No -> 3.

2. Do on-site levels exceed criteria by a substantial amount (over 10-100X)?
yes ->4.
no ->5.

3. Evaluate other possible uses, if all are likely to lead to less exposure than residential use - STOP, the site is unlikely to pose a risk and the no action alternative is appropriate. If other scenarios may be important, develop cleanup criteria using these scenarios and compare with on-site concentrations (step 1).
4. Either a restriction on use or remedial action is probably necessary. Is residential use an intended future use?
yes ->6.
no ->7.

5. Reevaluate exposure parameters used. Is use of actual site-specific values in place of default values likely to substantially affect risk calcs?
yes -> 8.
no -> 6.

6. Further action, either a refining of the risk assessment using site-specific exposure values, or remediation, or a combination of remediation and revisions to the risk assessment are necessary.

7. Develop criteria for maximum non-residential use scenario and compare with on-site concentrations (step 1).

8. Measure site-specific values and develop site-specific criteria using these values (step 1).

[a]Site can be used to refer to either the entire site or to a specific hot spot.

The re-focusing of risk assessment efforts involves both a removal or de-emphasis of certain parts of the current risk assessment and the addition or increased emphasis of other sections. A standard baseline risk assessment is outlined in Table 1. Sections of the risk assessment that can be de-emphasized if efforts are more focussed on cleanup goal determination are:

- Section 1 - Discussions of different potential future use scenarios,
- Section 4 - Evaluation of numerous exposure pathways, and
- Section 3, 5 - Evaluation of non-cancer risks associated with carcinogenic compounds.

The current approach to risk assessment is rather a shotgun approach to addressing site risks. Potential future site uses are established a priori and then risks posed by these hypothetical uses are evaluated. While considerable debate often takes place over the potential for the type of use to occur, once the potential uses are established (in Chapter One of the report), the hypothetical exposure scenarios generally carry through the entire assessment, in the process achieving some degree of authenticity. In order to ensure that human health and the environment are protected, if a particular use cannot be ruled out, it is generally included in the assessment. Consequently, future use scenarios that represent maximum possible use often drive the development of cleanup goals, even if such use is quite unlikely.

Put another way, selecting potential exposure scenarios in the first section of a baseline risk assessment means that cleanup levels developed based on the risk assessment must be designed to protect human health under a maximum use scenario (with associated large uncertainty). A more detailed reflection on the actual uses of the site (which decreases uncertainty) may result in remedial alternatives that are both health protective and substantially less costly. While this type of focussed assessment will not be possible in all cases, in situations where it can be used, the results will be much more realistic risk assessments. In essence, focussing the risk assessment on more accurate potential site uses is simply one approach to decreasing the uncertainty in the risk assessment process.

TABLE 1
Outline for a Baseline Risk Assessment

1.0 Introduction

This section outlines the purpose of the risk assessment and describes the procedures that will be used.

2.0 Site Characterization

Section 2 of the risk assessment provides background information on the site and characterizes the nature and extent of contamination. The selection of chemicals of potential concern is also discussed in this section.

3.0 Characteristics of Key Chemicals

In Section 3, information is presented on the environmental fate and transport of the key site contaminants and on the key toxicological characteristics of these compounds. As part of the toxicological assessment section, health-based criteria and media-specific guidance are also identified.

4.0 Exposure Assessment

An exposure assessment, including a discussion of exposure pathways, a description of potential receptors, and a quantitative estimate of exposure, is presented in Section 4. Modelling results are summarized in this section.

5.0 Risk Characterization

Section 5 contains the risk characterization, in which information on the toxicity of the chemical (from Section 3) is combined with the results of the exposure assessment (from Section 4) to determine if the site poses a risk to public health or the environment. Applicable or Relevant and Appropriate Requirements (ARARs) and a discussion of the uncertainties in the risk assessment are also included in this section.

6.0 Conclusions

The conclusions of the risk assessment and site-specific cleanup levels, are presented in Section 6.

7.0 References

One problem is that as soon as more realistic use scenarios are considered in the risk assessment, actual use will change to involve higher exposure (one of Murphy's Laws). Consequently, regulators, who are required to ensure public health, are forced to use maximum use scenarios, even if all parties agree such use is unlikely, unless some guarantee can be made as to a site's future use. Deed restrictions and limits on future land use are one way to ensure restrictions, but landowners are understandably reluctant to limit land use in the absence of clear evidence that such restrictions are essential. Consequently, risk assessments often evaluate a number of different scenarios and the particular scenario that will be used as the basis for establishing remedial objectives is often a major point of contention, with regulators requiring maximum use scenarios and property owners arguing that such scenarios are unreasonable.

One way around the concern about uncertain land use is to reformat the risk assessment. Focussing initially on maximum possible future use scenarios and only considering alternative uses during the final selection of remedial options can streamline the process. If the maximum use scenario does not indicate a risk, the evaluation is complete. Otherwise, the property owner needs to consider the desired and likely future use of the land to determine the most cost-effective combination of land use, remediation, and risk. More stringent cleanup goals would be required for scenarios that could lead to higher exposure (e.g., residential use) and less stringent goals would be necessary for lower exposure scenarios. This approach gives the land owner some control over the final outcome of the process and allows decisions to be made with a better understanding of the costs and benefits of specific actions.

For example, a land owner might be given the option of cleanup to residential remediation goals, at a cost of $3 million, or cleanup to less stringent goals with a deed restriction (that can be removed if further cleanup occurs or happens naturally) at a cost of $300,000. If the land is to be used for constructing fancy condominiums, the more expensive cleanup may be warranted. If the land is going to be used for commercial purposes, or is unused and the chemical degrades over time, then the less stringent criteria may be selected. In either case, the landowner has some information on which to base a decision.

In Section 4 of the standard risk assessment, a detailed and elaborate discussion of exposure pathways is generally prepared. However, sufficient information is available from numerous risk assessments prepared for Superfund sites and other hazardous waste sites so that the risk assessor can be pretty sure of the exposure routes that will drive the assessment. Exposure pathways can be limited to those that are known to drive risks at the particular type of site and for the particular chemicals detected. Finally, because of the method used to calculate cancer risks, for chemicals that are carcinogenic, their cancer risk will drive remediation, so a detailed discussion of noncancer risks for these chemicals provides little useful information. It may be possible to decrease other sections of the risk assessment (for example, detailed toxicity profiles may not be necessary for all chemicals), but these three sections (potential future use, exposure routes, and noncancer risks) represent some of the more contentious areas of most risk assessments. Discussing these issues only when they could actually impact the results of the assessment should decrease negotiation time and inherent costs.

Sections of the risk assessment that may require substantial upgrading include:

- Section 3 - Discussion of the environmental behavior of the chemicals of concern over time at the particular site, including a discussion of site-specific bioavailability,
- Section 3,5 - Discussions of uncertainties in the toxicity criteria,
- Sections 4,5 - Discussions of the uncertainties in the risk assessment including Monte Carlo simulation and qualitative evaluation (the risk manager making decisions on the site must have a range of options and a clear understanding of the options), and
- Section 6 - An overall discussion of the potential impact of particular cleanup values on remedial options.

More elaborate discussions on the transport and fate of chemicals in the particular environment of the site, including information on such factors as organic carbon content, porosity, metal speciation, and microbial populations may decrease uncertainty and allow less stringent criteria. For example, all other factors being equal, lower groundwater impacts may be associated with high lead levels in a limestone matrix than with lower lead levels in sandy soils with a low pH. Site-specific information on leachability and bioavailability are important components of this evaluation (3,124). Better defining of measurable exposure factors such as site-specific leachability or bioavailability can limit the need for conservative default assumptions. The use of measured, site-specific factors in place of standard default values may allow less stringent cleanup goals because of the decreased uncertainty in the risk assessment.

The importance of site-specific information on mobility cannot be emphasized enough. Default assumptions continue to be overused in risk assessment and in the resulting remediation. As an example, PCBs that have been present at a site for a short time and are associated with fuel oil are likely to be substantially more mobile than PCBs that were deposited 20 years ago and that are now highly weathered. The differences between these two situations is likely to have a substantial influence on risks and consequently should be considered in making remedial decisions. Better site-specific data may also be cost effective and necessary in light of an increased acceptance of non-fully destructive treatment technologies. For example, it may be possible to show through the use of a $10,000 treatability or leachability study that no migration of a constituent will occur at 1 ppm. This finding can then support the use of soil vapor extraction (at $80/ton) rather than excavation and incineration (at $500/ton) for a site containing 20,000 tons of chlorinated hydrocarbon-contaminated soils. The cost savings would obviously be substantial.

The toxicity criteria for chemicals were developed based on the best data identified at the time and in a small number of cases may be open to readjustment. For example, the reference dose for Mirex was revised upward (made less stringent) based on an updated toxicity assessment. If one or a very small number of chemicals are responsible for the majority of the risk at a site and if the chemical is not frequently encountered at hazardous waste sites, an evaluation of its toxicity may be useful.

As noted frequently in previous sections, a significant problem in risk assessment is the often unrealistic amount of over-conservatism that can be introduced into the risk assessment through the combining of several conservative assumptions. A much better quantitative evaluation of uncertainty is currently possible through the use of Monte Carlo simulation techniques (110). If a simplified risk assessment suggests that chemical concentrations at a site exceed allowable levels, it may be well worth performing such analyses to refine the risk assessment and remove some of the over-conservatism.

An annotated outline of a remediation-focused risk assessment is presented in Table 2. It is not possible to list all the facets of a risk assessment that might be changed to better characterize risk issues for the remediation because of the site-specific nature of the process. It is important to note that because the remediation will often affect the mobility of the chemical and consequently will affect the risks posed by the chemical, it may not be possible to develop final risk-based cleanup levels until after remediation is ongoing. Remediation goals should not be established as unchangeable values but as goals subject to the outcome of the remediation. Benzene at 1 mg/kg in sandy soils may pose a greater risk than benzene at 10 mg/kg in highly organic, very tight soils that have been treated to the fullest extent possible by soil vapor extraction.

Using risk assessment as a key component of the site remediation process should lead to lower ultimate costs for remediation and to a more rational cost allocation. Sites that are to be used for purposes that involve limited potential for exposure can be cleaned up to less stringent criteria than high exposure sites. The person likely to be impacted by the final decision (the land owner) can determine if they want a full cleanup at high initial cost or a combination of less stringent cleanup with some site control on future use (lower initial cost). Site-responsible parties are also likely to be willing to decrease uncertainty by taking site-specific measurements if they have some assurance that the associated expenditures will probably lead to higher cleanup goals and decreased remediation costs. Finally, the use of media-based cleanup goals as a point of comparison rather than an assessment of risk should have public perception benefits. The statement that PCBs in fish from a local pond pose less than a 10^{-6} cancer risk implies that someone is at risk of cancer. The statement that all PCB concentrations are below an allowable criteria of 2 mg/kg is less directly tied to personal risk (125).

TABLE 2

Summary of Changes to the Standard Risk Assessment

1.0 <u>Introduction</u>

This section should indicate that initially a single, maximum use scenario will be used in the assessment.

2.0 Site Characterization

Section 2 of the risk assessment should continue to provide pertinent information on the site and characterize the nature and extent of contamination.

3.0 Characteristics of Key Chemicals

In Section 3, additional information, perhaps based on site-specific treatability studies should be presented on the environmental behavior of the key site contaminants. For the toxicological information, only health-based criteria and media-specific guidance need to be identified, with a brief discussion of effects in the text and more information in an appendix.

4.0 Exposure Assessment/PRG Development

The exposure assessment should focus on key exposure pathways for the chemicals and situation encountered; a detailed evaluation of all possible pathways should be unnecessary. A better characterization of uncertainties, including perhaps, Monte Carlo analysis, may be useful. Rather than calculating an exposure dose, media-specific preliminary remediation goals are established based on the key exposure pathways.

5.0 Risk Characterization

Section 5 contains the comparison of the PRGs (and other media-specific criteria) with the concentrations of the chemicals at the site. For chemicals with site concentrations exceeding the PRGs, the areal extent should be characterized. A decision-tree approach such as that outlined in Figure 1 can be followed to decide the appropriate course of action at the site.

6.0 Conclusions

The conclusions of the risk comparison are presented in Section 6. A preliminary discussion of the potential impact of particular cleanup values on remedial options should also be included as part of this section. For example, the effect of bioremediation on the mobility of the non-degraded fraction of a target compound should be addressed.

7.0 References

3.0 DEVELOPING PRELIMINARY REMEDIATION GOALS

As an initial step in the risk assessment process as outlined above, maximum (generally non-site specific) exposure values need to be developed. Several states and EPA Region III have developed risk-based criteria for various environmental media (123,126). Preliminary Remediation Goals (PRGs) can also be developed following guidance presented in the EPA's (9) Risk Assessment Guidance for Superfund: Volume 1 - Human Health Evaluation Manual (Part B): Development of Risk-Based Preliminary Remediation Goals. These PRGs use maximum exposure scenarios and conservative assumptions so that it is unlikely that contact with the chemical-affected media of concern at the PRG levels would pose a significant health risk under almost any condition. The PRGs generally do not consider site-specific factors and depending on site characteristics and use, concentrations substantially greater than the PRGs may also not pose health risks. The PRGs are designed primarily to be used as a screening tool; if all levels of a chemical are below the PRGs, no further work is necessary. However, if the PRGs are exceeded, further analysis of the cleanup criteria including evaluation of site-specific factors is warranted and remediation *may* be needed.

3.1 Groundwater

The EPA (1991) Human Health Evaluation Manual Part B presents both detailed and summary (called reduced) equations for calculating PRGs for chemicals. The detailed equations are simply the equations presented in Chapter 4 with a set risk level (10^{-6}) or hazard index level (HI=1) and solved for chemical concentration. The reduced equations are useful for quickly calculating screening level criteria if no source of risk-based values is available. For calculating PRGs for noncarcinogenic chemicals in residential groundwater, the reduced equation is:

Eq (16) $$PRG = 73/[(7.5/RfD_i)+(2/RfD_o)]$$

Where, RfD_o and RfD_i are the oral reference doses for the chemicals.

For carcinogens in residential groundwater, the reduced equation is:

Eq (17) $$PRG = 1.7x10^{-4}/[(7.5xSF_i)+(2xSF_o)$$

Where, SF_o and SF_i are the cancer-slope factors for the chemicals of concern.

For both noncarcinogens and carcinogens, EPA (9) guidance for calculating PRGs recommends that if either an inhalation or oral RfD or slope factor is not available, this factor should be dropped from the equation. In addition, the PRG approach does not address dermal exposure and several studies noted by the EPA (27) suggest that dermal absorption of chemicals in water may be an important exposure pathway. In order to ensure that PRGs will not underestimate risk, for volatile and semi-volatile compounds (excludes the pesticides and metals) the PRG values derived for ingestion exposure alone are divided by a factor of two. This factor of two is based on the limited available data that suggests that inhalation exposure and dermal exposure pose risks roughly comparable (in some cases higher, in some cases lower, but in all cases the uncertainty is such that the range of possible outcomes encompasses the ingestion risk value) to that for ingestion exposure. The use of a factor of two rather than a factor of three is based on the consideration that semi-volatile compounds probably do not pose much of an inhalation risk and that for volatile compounds, it seems unlikely that an individual would be exposed via all three pathways on a regular basis. Using these values ensures that the PRGs achieve their intended purpose of acting as a lower bound on groundwater concentrations that could pose a health risk.

3.2 Soil and Sediment

The reduced equation for calculating PRGs for noncarcinogenic chemicals in residential soils is:

Eq (18) $$PRG = 2.7 \times 10^5\ (RfD_o)$$

Where, RfD_o is the oral reference dose for the chemical.

For carcinogens in residential soils, the reduced equation is:

Eq (19) $$PRG = 0.64/SF_o$$

Where SF_o is the cancer slope factor for the chemical.

For nonresidential (industrial/commercial) land use the reduced equations consider both direct ingestion and inhalation exposure. However, for persistent chemicals, volatilization is generally minimal except immediately after application and even for more volatile chemicals, their presence at the site for long periods of time probably substantially reduces the potential for volatilization. Eliminating the inhalation portion of the equation, the formula for noncarcinogens reduces to:

Eq (20) $$PRG = 2 \times 10^6 (RfD_o)$$

For carcinogens the equation is:

Eq (21) $$PRG = 5.8/SF_o$$

Per EPA guidance, the most stringent criteria (carcinogenic vs. noncarcinogenic) are used in establishing the values. However, as noted previously, because of the approach used to develop the toxicity criterion for carcinogens, cancer will always be the driving criteria for these compounds.

The values presented for noncarcinogens are generally developed for some extended period of exposure and typically use chronic toxicity values. This approach is generally sufficiently conservative to protect individuals who might be exposed for shorter time periods but several cases should be noted where this is not the case. Specifically, chemicals that are not particularly toxic under conditions involving long-term repeated exposure but that are acutely toxic may require supplemental short-term criteria. Three examples serve to illustrate this point. The soil criterion developed to protect against exposure via long-term, direct contact with soils for xylene, which is not particularly toxic, is 200,000 mg/kg, or 20%! Even short term contact with soil containing this level of xylene would certainly cause severe skin irritation. To prevent this type of problem, one rule of thumb (used by New Jersey (123) in its ISRA rules, for example) is to establish a generic maximum allowable soil concentration of 10,000 mg/kg or 1%, under the presumption that at higher levels, soil characteristics may be affected by the chemical or the chemical may pose acute health concerns.

As another example, ingestion of soil containing copper at the maximum daily soil ingestion rate estimated for children of 5 g/day (89) at a soil cleanup criteria of 600 mg/kg (based on long-term exposure) would lead to ingestion of 3 mg/day of copper. The toxicity criteria (minimum effect level or MEL) for copper (2.8 mg/day) is based on acute exposure and a single day of ingesting soil at a rate of 5 g/day would lead to exposure above this criteria. It should be noted that the MEL for copper was developed based on dissolved copper and that even exceeding the toxicity criteria might not lead to a manifestation of toxicity for copper in soils. Children probably get higher exposure to copper (with no apparent toxic effects) from eating pennies. However, it seems prudent to at least consider these effects. A similar and perhaps more convincing argument could be made for cyanide. Ingestion of 5 g of soil containing the long-term, risk-based level of 1600 mg/kg of cyanide would give a dose of 8 mg, or 0.5 mg/kg for a 15 kg child (126). This dose is too close to observed effect levels for cyanide of less than 1 mg/kg.

Short-term risks are generally less important than long-term risks in hazardous waste site risk assessment. However, as illustrated above, elevated intake levels that occur over short time periods need to be considered in assessing risks to ensure adequate human health protection. This issue is particularly important in that children are most likely to receive short-term exposure to elevated levels of these materials.

4.0 EXAMPLES

As noted above, risk assessment has primarily been used at Superfund and other hazardous waste sites to provide justification for enforcement actions and to determine the relative public health impacts of previously selected remedial alternatives. Only rarely is the risk assessor included in the selection of remedial alternatives for a site and it is even less likely that risk assessment will affect decisions on cleanup actions in any significant way. Increased discussion of the use of risk assessment is occurring and several states (e.g., New Jersey, Washington, Michigan) have taken the lead in developing risk-based cleanup criteria. However, the process of involving risk assessors in site remedial decisions has been surprisingly slow, in light of the fact that the remedial action was taken to protect human health and the environment in the first place. The resulting cleanup action may not adequately protect public health and/or may cost much more than necessary (sometimes millions of dollars more than necessary).

Risk Assessment is primarily used at hazardous waste sites to:

- Determine if a site poses a risk to public health or the environment,
- Compare remedial alternatives as to their relative risks, and
- Develop cleanup levels based on maximum exposure scenarios.

As noted above, risk assessment can and should be used to:

- Analyze Uncertainties--The risk assessor can assist the risk manager in making decisions regarding the effect of various assumptions on the estimation of "true" risks,
- Develop Appropriate Cleanup Levels--The risk assessor can use considerations of site-specific factors to develop cleanup levels that are appropriate for a particular site, and
- Develop Appropriate Cleanup Remedies--Interactions among risk managers, remediation experts, and risk assessors can lead to the selection of more efficacious remedies.

4.1 Analysis of the Effects of Uncertainty in the Assessment

Use of standard approaches to quantitative risk assessment such as those presented in the EPA (3) RAGS Manual may produce results that suggest a significant risk is present. However, consideration of other factors that are difficult or impossible to quantify may show clearly that the actual risk is likely to be much lower. EPA notes in their guidance that a subjective consideration of site-specific factors is necessary but this is sometimes ignored. Failure to include the risk assessor in the process of selecting cleanup remedies at sites where such poorly quantifiable factors are present may lead to an unnecessarily stringent cleanup with consequently excessive expenditures of personnel and funds.

An example of the problems that rely on a strict interpretation of the baseline risk assessment is provided by a site in upstate Wisconsin. The site is located in a rural wetland area approximately 500 yards from a large lake. The site had been used as a dumpsite for many years.

PCB-contaminated demolition material was disposed at the site in 1978. This material was removed and taken to a hazardous waste landfill. Subsequent sampling of the area revealed the presence of a white clay-like material containing elevated levels of lead and chromium, 2 - 3 feet beneath the surface to the west of the PCB-contaminated area.

An evaluation of the nature and extent of contamination and of potential exposure to this material revealed that lead was present at an average concentration of 170 mg/kg and a maximum concentration of 1200 mg/kg. Chromium was present at a mean of 130 mg/kg and a maximum of 500 mg/kg. No information was available as to speciation of the chromium, an important factor, as Cr VI is much more toxic and mobile than Cr III. Leachability studies indicated that the metals were tightly bound to the soil (suggesting that CrVI was probably not present). Contaminated soils are generally more than 1 foot below the surface. The site is covered with trees and other vegetation and is in a marshy location with very shallow groundwater which would prevent the construction of houses in the location without importing fill (i.e., without filling a wetlands).

The baseline risk assessment indicated that under a worst case exposure scenario involving the construction of a residential dwelling directly on the site, lead levels would exceed allowable limits and a child playing outdoors in contaminated soil almost daily would have a slightly elevated risk of adverse health effects as a result of lead exposure. However, the assessment noted that such exposure was extremely unlikely because: 1) building a house on the site would be difficult because of the low-water table, and 2) even if a house were constructed, it is unlikely that a child would be exposed to contaminated soil that is located a foot or more beneath a ground surface that is likely to be covered by grass or other vegetation.

The remedial action selected for the site consisted of covering a 5 acre area with a protective clay cap and monitoring groundwater. The total cost of this alternative is expected to exceed $500,000. In this case, the risk assessment was only used to answer the question: Could the site pose a public health risk? Once it was established that a risk could exist (even though it was heavily qualified and noted to be highly unlikely to exist) selection of a remedial alternative proceeded with little consideration of public health and environmental concerns. (Indeed, removing 5 acres of wildlife habitat to install a clay cap may have an adverse environmental impact.)

The baseline assessment concluded that cleanup of the metals at the site was unnecessary because no exposure to these metals was likely and it appeared that migration was also unlikely. At most, cleanup of a few hotspots of contamination might have been warranted to ensure that even under a worst case exposure scenario the site did not pose a public health risk. Consideration of the risk assessment in more detail may have led to savings of up to the $500,000 cost of remediation if as seems likely, this consideration indicated that the no-action alternative was appropriate. Risks were only posed by an exposure scenario involving direct and frequent contact with hotspots of contamination. Cleanup of only these hotspots would have saved several hundred thousand dollars.

4.2 More Appropriate Cleanup Goals

Remedial actions at hazardous waste sites often involve cleanup to generic standards such as the EPA drinking water standards (MCLs) or other ARARS. However, use of these generic standards as cleanup goals may lead to excessive cleanups that have little or no impact on risk levels. Detailed consideration of factors such as site conditions, uncertainties in the risk assessment, properties of the chemicals, and potential human exposure may allow the risk assessor to significantly revise these cleanup numbers.

Risk assessment can be used as a vehicle to present site factors that can serve to decrease risk and consequently as a way to increase allowable levels of residual chemicals in environmental media. An example is provided by a site located in a rural area in Northern California in an area of limited industrial use. The site consists of a vacant lot that is used as a storage area for a neighboring truck repair shop. The lot is very sparsely vegetated and soil in the lot consists of a sandy loam. Groundwater is about 15 feet below ground surface and discharges into a major river that is located approximately 500 yards from the site. The entire site is fenced. Temperatures in the summer tend to exceed 90°F.

During the summer, a tank truck accidently spilled approximately 2000 gallons of fuel oil on the vacant lot. Because of the nature of the soils, this oil soaked into the ground almost immediately. Although excavation of visibly contaminated soils was started within 24 hours, sampling at the site one week after the spill revealed the presence of total petroleum hydrocarbons (TPH) at levels around 100 ppm.

Removal of the material to a hazardous waste landfill was selected as the remedial action. Problems were encountered when a state agency determined that the site had to be cleaned to a TPH level of 10 ppm, an order of magnitude below common cleanup goals. Preliminary sampling results and consideration of the nature of the site (truck maintenance yard) suggested that cleanup to 10 ppm would almost certainly have involved removal of much more than the 2000 cubic yards of soil and rocks already removed.

The state lowered the TPH cleanup levels on the grounds that the spill had occurred on sandy soil near a major river that contained a sensitive species (Salmonids). However, the judgement failed to consider the following countervailing factors: 1. Fuel oil consists of a mixture of hydrocarbons with varying molecular weights and consequently varying solubilities and varying propensities to bind to soil particles, 2. Contaminated soils in the excavation had been exposed to temperatures above 90°F for over a week and the more volatile (and more soluble) components had almost certainly evaporated, 3. The fuel oil fraction that remained on site would be rather tightly bound to soil particles, and 4. Mass balance computations suggested that only small quantities of oil remained on site. Further discussion of these issues with state personnel led to the approval of the remedial action with no additional excavation required.

In this case, use of risk assessment prevented additional soil removal, disposal, and sampling, and provided a cost savings of many thousands of dollars. More extensive use of risk assessment early in the project may have allowed even more lightly contaminated soil to remain on site. Other treatment technologies were not used because previous experience had shown that it was difficult to reduce TPH to below 100 ppm. A risk assessment that showed that higher levels of TPH could have been left and still ensure that no leaching into the river occurred may have allowed use of these alternative treatment technologies.

4.3 More Appropriate Cleanup Remedies

Risk assessors are often presented with several remedial alternatives and requested to evaluate the risks of each, relative to the no-action (baseline) alternative. However, it is often clear that a combination of parts of several alternatives may provide the same level of protection at greatly reduced costs or that an alternative that could not meet maximum exposure criteria would still be sufficiently health protective. Interaction among the risk assessor, the remedial contractor, and the risk manager ensures that the most efficacious remedial alternative is selected.

The site is located in a suburban area in southern California that is zoned for mixed industrial and residential use. Most of the residents obtain their drinking water from a municipal well but several private wells are used in the area. The nearest private well is located about a mile downgradient of the site.

A manufacturing facility was operated on the site for a number of years. During this time, waste solvents were disposed of by pouring the material into a drain leading to a backyard sump. Sampling revealed the presence of trichloroethylene (TCE) and tetrachloroethylene (PCE) in soils beneath much of the site.

Over 5000 cubic yards of soil were contaminated with elevated levels of TCE or PCE and groundwater beneath the site contained levels exceeding the MCL of 5 ppb. Modelling results indicated that the contamination would likely reach the downgradient water supply well in about 10 years if no remedial action were taken. Levels at the downgradient well were estimated to be present for 20 years at an average concentration over this time of 50 ppb and a maximum concentration of 500 ppb for a one year average.

Four remedial alternatives were considered for the site: No Action, Groundwater Pump and Treat with Air Stripping, Installation of a Clay Cap, and Excavation. The results of the baseline risk assessment indicated that the no-action alternative posed a public health risk and it was rejected for this reason. Installation of a clay cap was also considered unlikely to adequately prevent migration of contaminant to the drinking water well and it was also rejected. Pumping and treating the water was considered likely to effectively remove the contamination but there was a substantial possibility that this technique would lead to subsidence and that it would create numerous associated problems. Excavation was the most costly remedy but would clearly protect public health in the area and would not cause other adverse effects. Therefore this alternative was selected as the remedial action. Total cost for the cleanup was anticipated to exceed $2,500,000.

The risk assessor was presented with the four alternatives and asked to determine the relative effect of each on public health. The assessment revealed that two of these protected public health but only one was practical from an engineering standpoint. Thus, by default, the most expensive measure was selected.

Much of the contamination (over 75% of the mass of the TCE and PCE) was present in the area of the sump. Modelling results showed that if the 500 cubic yards of soil located in this area were remediated either by excavation or by a technique such as *in situ* soil vapor extraction (SVE), only low levels of TCE or PCE would reach the drinking water well. Furthermore, the addition of a small extraction well and air stripper would remove sufficient contamination to prevent exposure but at the same time would not cause subsidence. Interaction between the risk assessor and the remedial contractor could have led to the selection of this integrated approach at a substantial savings.

5.0 PERSPECTIVE

Increased involvement of the risk assessor in the selection of remedial alternatives can lead to more efficient use of limited funds and personnel. This will free these resources for work on other sites. Risk managers have, in general, gotten much more knowledgeable about the risk assessment process and are much more likely to consider risk levels in selecting remedial alternatives. However, there is room for improvement, as full involvement of a risk assessment professional in the selection of remedial alternatives is still uncommon. As illustrated by these examples, more involvement of the risk assessor in the remedial process can lead to more effective expenditure of limited funds. In addition, it can prevent excessive work on a site with low risks, which can divert limited personnel from other, potentially more hazardous sites.

6.0 GENERAL REFERENCES

Smith, R.L., Risk-Based Concentration Table, Third Quarter 1993, Memorandum from R.L. Smith, Senior Toxicologist to RBC Table mailing list, U.S.EPA, Region III, Philadelphia, PA July (1993)

Environmental Protection Agency (EPA), Human Health Evaluation Manual, Part B: Development of Risk-based Preliminary Remediation Goals. (Interim) Washington, D.C., OSWER Directive 9285.7-01B. (1991)

9.

RADIATION RISK ASSESSMENT

This chapter presents a brief summary of risk assessment procedures at hazardous waste sites where ionizing radiation is a concern. The approach to radiation risk assessment has some similarities to that for chemical risk assessment and many of the same characteristics are important (e.g., environmental behavior of the element at the site, location of potential receptor populations). Radiation risk assessment differs in dosimetry (the type of and measurement of dose received) and has an increased layer of complexity in that the dose to individual target organs is evaluated rather than simply the whole body dose. Radiation risk assessment is complex and substantially different from chemical risk assessment, and the purpose of this chapter is only to provide an introduction to the subject. One type of radiation risk assessment, the assessment of risks associated with the release of radon gas, is fairly similar to chemical risk assessment and approaches to evaluating health risks for radon are presented in some detail as an introduction to health assessment for radionuclides.

1.0 DOSIMETRY

Chemical dose is measured in terms of milligrams of chemical. For radiation, dose is measured in terms of amount of ionizing energy released. Sources of this energy include alpha and beta particles, and gamma radiation (127). Alpha particles consist of two protons and two neutrons and have a +2 charge. Because of their relatively large size, these particles have a poor ability to penetrate and tend to release their energy over a short distance. Alpha particles only affect cells in the immediate vicinity of their release or on exposed surfaces, and are of most concern for unprotected epithelial cells such as those in the lungs. Gamma radiation on the other hand, has high penetrating power and can have effects on tissues throughout the body. Alpha particles transfer considerably more ionizing energy over a given distance in a tissue than does gamma radiation. This difference in linear energy transfer (LET) causes alpha and gamma radiation to differ substantially in their Relative Biological Effectiveness (RBE). That is, a given absorbed dose (measured in units called rads) of high-LET alpha energy is much more likely to adversely affect a cell than the same absorbed dose (in rads) of low-LET gamma radiation. This difference can be corrected by using units called rems, which are equal to the absorbed dose in rads multiplied by a quality factor, which adjusts for the difference in RBE. Alpha particles are generally considered to be over 20 times more biologically effective than gamma radiation. However, gamma radiation is also of concern because it's high penetrating ability can affect a large number of target tissues.

Allowable levels of radiation have changed significantly over the years. Early regulations were focused on protecting individuals from direct toxicity to the radiation. With an increasing awareness of the potential for radiation to cause genetic damage in both developing fetuses and in somatic cells (i.e., cancer), regulations have focussed more heavily on protecting these sensitive endpoints. Table 1 presents a very brief summary of the regulatory history for radiation. Current regulations are changing somewhat but overall the principle of limiting exposure to As Low As Reasonably Achievable (the principle of ALARA) continues to be the primary regulation for radionuclides.

2.0 RADON

Thorium-230 degrades with a half-life of 80,000 years to form radium-226 (Ra-226), which degrades with a half-life of 1,600 years to form radon-222 (Rn-222), a noble gas (128). Radon-222 and the radionuclides formed by subsequent decay, called radon daughters or radon progeny, undergo rather rapid decay.

TABLE 1

Regulatory History for Radionuclides

Year	Standard	Basis
1920	5R/month, 0.2R/day	Tolerance dose--fraction of the dose that produces erythema
1934	0.2 R/day	Tolerance dose--no harmful effects in radiologists at this level
1936	0.1 R/day	Based on potential for genetic effects
1949	0.3 rem/week	U.S. National Committee on Radiation Protection (NCRP)
1959	170 mR/yr (each) from (1) medical and (2) other man-made sources	Standard to protect child-bearing generation from genetic damage
1971	5 rem/year (occupational) 0.5 rem/yr (non-occupational) 0.1 rem/yr (student) 0.17 rem/yr (population)	NCRP
1978 be	ALARA	NCRP; USEPA --limits should set as low as reasonably achievable (ALARA)

R=roentgen, a measure of exposure. For practical purposes, 1R = 1 rad.

Radon gas and its daughters are released into the ambient air following the decay of radium. In the external environment these particles diffuse fairly rapidly, but they can accumulate in the indoor environment (129,130). Individuals living or working in buildings with significant levels of radon and its progeny may suffer an increased risk of lung cancer due to inhalation of the radionuclides and subsequent internal release of radiation. Gamma radiation emanating from radium-contaminated soil may also pose a health risk to residents of a contaminated area.

The short-lived decay products of radon-222 emit alpha and beta particles and gamma radiation. Alpha particles, which have a large mass and consequently high ionizing energy, have the greatest adverse effect on cells. However, as noted above, alpha and beta particles have limited ability to penetrate tissue. Therefore, they are most dangerous to unprotected epithelial tissue such as that lining the lung and gut.

Gamma radiation will also affect these tissues but has greater ability to penetrate tissue and has been associated with an increased incidence of cancer at several sites (130). It is noted here only as a harmful component of radon decay, which will add slightly to the risk associated with the alpha particle, but will be discussed more fully in a separate section. Radon and its daughters may be absorbed by the body from either the lung or gut, but the risk of cancer from an absorbed dose is probably insignificant relative to the risk of lung cancer.

2.1 Radon Dosimetry

The concentration of radon progeny in the air is measured in units called working levels (WL) and cumulative exposure to radon is measured in units of working level months or WLM (128). These units are comparable in terms of their use, to mg/m^3 of chemical for the working level and mg/kg/day for the working level month. A working level is a standard measure of radon daughter concentration in air and is equal to any combination of radon daughters that produces 1.3×10^5 Mev (million electron volts) of alpha energy per liter of air in decaying through Polonium-214. A working level month is a measure of cumulative exposure which is equal to an exposure of 1 WL for 173 hours or for roughly 170 hours (128, 131). Other equivalencies are shown in Table 2. Currently the Mine Safety and Health Administration (MSHA) of the

Department of Labor, limits the maximum permissible concentration of radon daughters in mines to 1.0 working level (WL) and limits occupational exposure to 4.0 working level months (WLM) over a calendar year (131). For exposure to the general population, the U.S. Environmental Protection Agency (USEPA) has determined that in any occupied or habitable building, the concentration of radon decay products (including background) shall in no case exceed 0.03 WL, and that a reasonable effort should be made to achieve an annual average radon decay product concentration (including background) of 0.02 WL (40 CFR 192.12).

Numerous groups have reported on the excess risk associated with exposure to radon and its decay products (128,131). All of the researchers used data from the exposure of uranium mine workers to radon daughters and linear extrapolation to determine the potential risk to either the general population or to persons exposed to high levels of radon in their homes. Data from Europe and Canada gave higher relative risks than data from the United States and therefore the risk assessments prepared by the different researchers varied by as much as an order of magnitude.

The available studies suggest that the excess risk associated with exposure to radon is between 10^{-3} and 10^{-4} per WLM of exposure. This amount of difference is probably within the range of error for the studies. Smokers have a higher relative risk than non-smokers, and there is some evidence that smoking and radon may act synergistically to increase the risk of lung cancer. Table 3 provides some data collected in homes built on fill material from radium watch factories and provide some indication of the magnitude of the risks associated with exposure to radon. The EPA risk estimates suggest that a lifetime (70 year) exposure to indoor air concentrations of radon of 0.02 WL (the EPA target level) will produce a risk of about 0.2-2% of developing a fatal lung cancer. Linear extrapolation from these estimates suggests that exposure to the background level of radon (0.004 WL) is associated with a fatal lung cancer risk of between 4×10^{-4} and 4×10^{-3}. Most regulatory agencies use a risk value of 10^{-6} as a target risk level for chemical carcinogens and most federal regulations for environmental contaminants (incorporating consideration of costs and feasibility) have generally fallen into the 10^{-4} to 10^{-6} lifetime risk range. Because these ranges are unattainable for this situation, radon emissions are generally reduced to the lowest feasibly achievable level also known as As Low As Reasonably Achievable or ALARA.

TABLE 2

Conversion Factors and Dose Equivalents for Radon-222 and Its Progeny

Concentration of Radon-222	Working Level	Dose Equivalent
Continuous[a] exposure to 1 pCi/m^3	Continuous exposure to 5 x 10^{-6} WL	4 mrem/ year to the lung[b]
Continuous[a] exposure to 2 x 10^5 pCi/m^3	Continuous exposure to 1 WL	8 x 10^5 mrem/year to the lung
Continuous[a] exposure to 0.25 pCi/m^3	Continuous exposure to 1.3 x 10^{-6} WL	1 mrem/ year to the lung

[a]Assumed radon-222 and daughter ratios 1.0/0.9/0.5/0.35 (expected ratio for living accommodation with normal ventilation).

[b]Dose delivered to basal cell nuclei of segmental bronchi, quality factor of 10 for alpha particles, penetration depth to cells at risk, 60 um.

TABLE 3

Estimated Risk[a] of Lung Cancer Associated with Exposure to Radon-222 and Its Progeny in Montclair, West Orange, and Glen Ridge, New Jersey and in the "Average" New Jersey House

Concentrations (WL)	Tier	Total WLM	Risk
>0.5	A	1,125	$>5x10^{-1}$
0.1-0.05	B	225-1,125	$9x10^{-2}$ to $5x10^{-1}$
0.02-0.1	C	45-225	$2x10^{-2}$ to $9x10^{-2}$
0.007-0.02	D+	15-45	$6x10^{-3}$ to $2x10^{-2}$
0.002	D, average background	4.5	$2x10^{-3}$

[a]Calculation based on EPA (USEPA 1979) and NCRP (1984) figures; see text for explanation.

3.0 GAMMA RADIATION

Gamma radiation is highly penetrating and causes a relatively uniform exposure to the body (127). Because of this, it is likely to increase the incidence of tumors at all sites, although certain cells or tissues are more sensitive than others and more likely to be adversely affected, specifically, those tissues with rapidly dividing cells. Tumors most commonly associated with gamma irradiation are leukemias, lymphomas, and cancer of the breast, lung, esophagus, stomach, and urinary tract. However, most reviews consider mortality from all cancers when assessing the risk associated with gamma radiation.

In addition to causing cancer, gamma irradiation also produces genetic disorders by causing chromosomal aberrations and gene mutations in germ cells. The genetic disorders observed following exposure of mice to ionizing radiation were manifested as spontaneous abortions, and behavioral and physical abnormalities.

The Department of Energy has established Standards for Radiation Protection of the Public of 500 mrem/year and 170 mrem/year for maximum dose to an individual and average dose to a sample of an exposed population, respectively, for exposure to the whole body, gonads, or bone marrow. EPA regulations limit gamma radiation to a level of 20 μroentgens/hour (equal to 20 μrads/hour) above background in occupied or habitable buildings (40 CFR 192.12).

The EPA (131) and others have calculated the risk of fatal cancer for persons exposed for their lifetime to 100 mrem/year of external penetrating radiation. Risks were based on both absolute and relative risk models and linear extrapolation from high doses to low doses. The reported values for risk at 100 mrem/year were converted by linear extrapolation to the risks associated with exposure to 500 mrem/year for a lifetime (the radiation protection standards). The risks predicted by each model are presented in Table 4. As can be seen from this table, target risk levels for gamma radiation, as for radon, are well above the range of 10^{-4} to 10^{-6} commonly used in regulating chemicals. Again, the principle of ALARA is applicable to establishing radiation standards.

4.0 PERSPECTIVE

This chapter at best only briefly touches on a very complicated topic. There are certainly similarities between radiation risk assessment and chemical risk assessment but in many ways they are very different. Because of these differences, mixed waste, the combination of low level rad-wast and chemical wastes present some rather interesting issues. For example, benzene in air at a level posing a 10^{-4} risk is above the allowable chemical risk-based target level, while tritium at a risk of 10^{-3} is below the target level. Is it more important to control risks (in which case tritium should be controlled, even though it is below its standard) or

TABLE 4

Estimated Lifetime Risk of Excess Fatal Cancer Associated with Lifetime Exposure to 500 mrem of Highly Penetrating Radiation Annually

Source	Relative Risk	Absolute Risk
EPA 1979	2.4×10^{-2}	4.2×10^{-3}
BEIR III 1980[a]	6×10^{-3}	2.5×10^{-3}
BEIR III 1980[b]	1.5×10^{-2}	5×10^{-3}
Radford	2.7×10^{-2}	6×10^{-3}

[a]Linear-Quadratic Extrapolation
[b]Simple Linear Extrapolation

Source: (131,132)

to control to set standards (in which case benzene would be controlled)? These types of issues make the mixed waste situation particularly interesting.

A final note on radiation risk assessment is that as for all other types of risk assessment, a considerable amount of uncertainty is inherent in the process. One criticism of radiation risk assessment is that in many cases, more significant figures are presented than are warranted by the data. Radiation dosimetry tends to be based on better information than cancer risk data but a considerable amount of uncertainty still exists, and radiation exposure calculations are essentially equal in uncertainty to the same calculations performed for chemical exposure.

5.0 GENERAL REFERENCES

Committee on the Biological Effects of Ionizing Radiation (BEIR III). 1980. *The Effects on Populations of Exposure to Low Levels of Ionizing Radiation:* National Academy Press, Washington, D.C. (1980)

Environmental Protection Agency (EPA). 1983. *Background Information Document--Proposed Standards for Radionuclides.* U.S. Office of Radiation Programs, Washington, D.C., EPA 520/1-83-001 March (1983)

Hobbs, C.H. and McClellan, R.O., Radiation and Radioactive Materials. In: *Casarett and Doull's Toxicology--The Basic Science of Poisons.* (Doull, J., Klaassen, C., Amdur, M., eds.) 2nd ed. Macmillan Publishing Co., New York (1986)

National Conference on Radiation Protection and Measurement (NCRP), *Evaluation of Occupational and Environmental Exposures to Radon and Radon Daughters in the United States.* NCRP Report No. 78, Bethesda, MD (1984)

Shapiro, J., *Radiation Protection: A Guide for Scientists and Physicians,* Second Edition, Harvard University Press, Cambridge, MA (1981)

10.

ECOLOGICAL RISK ASSESSMENT

The primary purpose of site-specific ecological risk assessment is to evaluate whether or not ecological receptors (i.e., wildlife) have been adversely affected by a hazardous waste site. The ecological risk assessment can also be used to determine if planned remediation activities will cause an ecological impact, and to develop appropriate remedial goals/options for a site. The ecological assessment can be either simple or very complex depending on factors such as site chemicals, flora and fauna, and location of the site. This chapter is designed to provide a very brief overview of this subject. More thorough reviews of the topic are available in references listed at the end of the chapter.

1.0 INTRODUCTION

Ecological risk assessments conducted for hazardous waste sites in the past have commonly contained a very limited amount of information. In many cases, the assessment consisted of only a listing of endangered species that potentially could be found in the area (in one recently reviewed risk assessment, the list consisted of all endangered species in South Carolina, which included several whales; rather unlikely to be found near an inland site), a list of flora and fauna observed on the site, and a brief discussion of the potential for site chemicals to enter surface water bodies. These assessments were of limited usefulness although they potentially could identify major problems (e.g., hexachlorobenzene in a marsh might be noted as a risk to fish, predatory birds, and mink although no discussion of its effect on the microenvironment would be noted). Recently, ecological assessments have contained a greater level of detail and in some cases have included very thorough quantitative evaluations of the effects of site-specific chemicals on the ecological community.

The level of detail required for an ecological risk assessment, as for a human health assessment, is dependent on the types of chemicals detected at the site, the potential migration pathways for these chemicals, and on site characteristics such as location. An evaluation of the types of habitat present and a discussion of the types of plant and animal life that have been observed or that could be present in particular habitat types is conducted. Particular emphasis is placed on threatened or endangered species that may be present on the site and depending on the type of investigatory or remedial activities that will be conducted, detailed surveys of specific areas of the site may be required.

2.0 APPROACH AND FORMAT

For ecological risk assessments, guidance is presented in the EPA Risk Assessment Guidance for Superfund Volume II: Environmental Evaluation Manual and in subsequent support documents (133,134,135). However, the guidance presented on environmental risk assessment is quite subjective and unlike for human health risk assessment, does not clearly outline the approach that should be used. This lack of clear guidance is in part a function of the complexity of the system being evaluated and the large number of endpoints that must be considered. For a human health assessment, the risk assessor has only to worry about whether or not a particular chemical or chemicals will adversely affect a single species, humans. The risk assessor has at least some information on toxicity to humans or surrogate species, data on intake levels, and some understanding of likely human behavior. In the case of an ecological assessment, a huge range of organisms (including both plants and animals) may be affected, data on the toxicity of chemicals to each of these species or to appropriate surrogates may not be available, and the potential for exposure can vary widely. In addition, the amount of interdependence in the ecosystem may serve to increase the impact of a toxicant with apparently limited effects, while the effects of tolerance or avoidance behavior can decrease chemical effects (136). Because of the variation and increased complexities, a critical component of an ecological risk assessment is a detailed discussion of the approaches that will be used and its goals and objectives.

A generalized outline for an ecological risk assessment is provided in Table 1. The first section of an ecological risk assessment describes its goals and objectives. As for the human health risk assessment, the primary objective is to determine if chemicals at the site pose a risk to potential receptors. Secondary functions can include an evaluation of the effect of planned remedial options and development of site-specific cleanup goals. An ecological evaluation can range from a simple (strictly qualitative evaluation) to a very detailed and complex assessment including bioconcentration studies, an evaluation of species diversity, floral and fauna surveys, etc. Section 2, on site characterization, includes a discussion of the nature and extent of contamination, a description of habitats, and a discussion of organisms present or likely to be present at the site. The site characterization section of an eco-risk assessment is generally larger and more complex than the same section of a human health risk assessment. Ecosystems and habitats need to be described, the species present noted, and any known or possible threatened or endangered species noted.

Key characteristics that may have an impact on the ecology of the area, and species or habitats that could be adversely affected are identified in Section 3. This section also contains a discussion of threatened or endangered wildlife that could reasonably be expected to use the site or surrounding areas. Any potential threat to these species posed by remedial activities at the facility should also be discussed. The toxicity section provides available information on toxicity criteria for the chemicals of concern (EPA Water Quality Criteria and other available values) and other information on chemical toxicity.

TABLE 1

Outline of an Ecological Risk Assessment

1. INTRODUCTION
2. SITE CHARACTERIZATION
    2.1 General Site Characteristics
    2.2 Nature and Extent of Agent of Concern
    2.3 Habitats on the Site
    2.4 Flora
    2.5 Fauna
    2.6 Threatened or Endangered Species
3. CHARACTERISTICS OF AGENTS OF CONCERN
    3.1 Environmental Behavior
    3.2 Hazard Evaluation
4. EXPOSURE ASSESSMENT
    4.1 Exposure Routes
    4.2 Target Organisms
    4.3 Evaluation of Potential Exposure
    4.4 Quantitative Exposure Assessment
5. RISK CHARACTERIZATION
    5.1 Estimated Risks for Single Endpoints
    5.2 Multiple Chemical/Species Risk Evaluation
    5.3 Risks to Communities and Ecosystems
    5.4 Uncertainty
6. CONCLUSIONS

In Section 4, potential exposure routes, chemical behavior in the environment, and an evaluation of possible exposure are presented. Exposure routes that may be addressed for the site include but are not limited to, release of chemicals into surface water bodies (effects on aquatic and terrestrial organisms using the water), direct contact of plants or burrowing animals with contaminated soils, and exposure routes involving biomagnification through the food web. Any facility-specific toxicity testing should be described in this section of the report. The final section of the report, the risk characterization section, describes to the extent possible, the potential for chemicals at the landfill to adversely affect individual species or ecosystems. Again, as for the HRA, a discussion of the uncertainty in the assessment is a critical part of the ERA.

2.1 Literature Review.

A comprehensive literature review, covering the general ecology of the area, population and community studies, chemicals of concern, and sensitivities of potential indicator species, and other pertinent subjects is generally the first step. Data generated during the literature search provides background information on relative abundance and kinds of organisms present, whether these organisms have exhibited sensitivity to potential chemicals of concern, and general background ecological information that will aid in characterizing conditions at the site. A good source of information is local universities. Also, many towns may have been required to perform evaluations on local waterbodies and these studies may supply useful data.

2.2 Screening Studies

A large number of screening tests have been developed that are designed to allow rapid evaluation of site conditions. These test systems have been designed for aquatic and terrestrial systems and include both laboratory and field tests. For laboratory tests, the environmental media are collected and diluted with clean media to known percentages (i.e., the test substrate will contain between 0% and 100% of the site media). The percent survival of the test organism in the system is then observed. Test organisms include the earthworm, various plants (germination), minnows (commonly the fathead minnow, *Pimephales promelas*), and several aquatic invertebrates.

Field surveys and field testing involving the introduction of test organisms into both background and on-site environments are also conducted to evaluate the potential for adverse impacts on the ecosystem. Most of these test methods are designed to evaluate effects on aquatic systems. However, because site chemicals will tend to migrate towards drainageways, procedures that evaluate impacts on stream communities can also provide information on the presence of mobile constituents in terrestrial soils. As an example, EPA's Rapid Bioassessment Protocols (RBPs) for benthic organisms and fish can be used to evaluate effects on flowing water (stream and river; also called lotic) communities (137).

The RBP approach has several levels, with the first level (RBP I for benthic organisms and RBP IV for fish) being primarily screening tools and subsequent methods becoming increasingly quantitative in nature (136). RBP I, for example is designed to allow a trained biologist to distinguish between clearly impacted and non-impacted areas and areas that require further study. The method is based on the best professional judgement of the biologist and an assessment of the types of organisms present in the stream versus the types that would be expected given the habitat and physical parameters. Physical parameters such as dissolved oxygen, temperature, pH, hardness, and conductivity are measured in the field at each RBP sample site and used in conjunction with the assessment of organisms to assess the results. Certain organisms are more or less pollutant sensitive and their presence or absence can serve as an indication of a healthy or stressed environment. For example, the presence of pollution sensitive Ephemeroptera, Plecoptera, and Trichoptera (EPT) organisms is generally indicative of a healthy environment, while their absence and the presence of more pollutant tolerant species such as Chironomidae would be cause for concern. The presence or absence of species may be related to natural factors and the measurement of physical parameters can assist in judging whether a low number of benthic organisms is site chemical-related or caused by habitat differences.

2.3 Wetland Delineation

Wetland delineation is conducted in order to provide guidance for remediation (i.e., to ensure that remedial actions do not adversely affect wetlands). Guidance for conducting wetland delineation is contained in the Federal Manual for Identifying and Delineating Jurisdictional Wetlands (138). This manual includes approaches that can be based on literature/map review and field approaches.

The literature review approach to wetlands classification involves locating the area on a topographic map (USGS) and delineating the approximate boundary of the area under consideration on the map. Once this area is located, other sources of information on wetlands are reviewed, including National Wetland Inventory (NWI) maps, or State or local wetland maps. If Soil Conservation Service maps are available, these are also reviewed. Finally, aerial photographs, any site-specific data (studies by ecologists at local universities, for example) and other information that may be appropriate, such as studies of similar areas by State ecologists should be reviewed.

Wetlands are determined to be present if the area under consideration has been shown to be a wetland on a NWI or other wetland map, or if hydric (water saturated) soil shows on a soil map. A site may also be considered a wetland if site-specific information shows hydrophytic vegetation, hydric soils, and/or wetland hydrology, or if signs of wetlands appear on aerial photos. Finally, combinations of these factors are important in reaching a definitive conclusion regarding the site.

An on-site wetlands delineation consists of determining whether or not an area should be considered a wetlands based on an evaluation of site characteristics in the field. As a first step, the approximate limits of the area of concern are identified and local area that may meet the criteria for wetlands are outlined on a topographical map or aerial photograph (e.g., lowland areas might be outlined, while hilltops might be removed from consideration). Once the areas for review have been determined, these areas should be inspected to determine if obvious signs of wetland hydrology are present. For areas that are not clearly designated based on a preliminary scan, consideration of such factors as whether the current environmental conditions are indicative of normal conditions, whether seasonal or longer-term fluctuations may occur, and whether appropriate plant species are present may help in making a determination. It may not always be possible to firmly delineate wetland areas, because of fluctuations in systems over time.

2.4 Terrestrial Faunal Surveys.

Surveys of terrestrial organisms can be used to establish whether levels of the organisms are different than expected and to determine if all expected species are present. The lack of a normally common species may indicate that this species is under stress in the area. Surveys can be conducted visually (e.g., by counting insects, insect damage, or birds in a particular area, or by trapping small animals, such as voles) (139, 140, 141). The results of site surveys can be compared with literature values to make a qualitative assessment of whether or not a decline or impairment in these communities has occurred that could be related to the hazardous-waste site. In general, particular attention is given to State or Federal Threatened and Endangered species. Threatened and endangered species are of concern in both ensuring that they are not affected by the site in its current condition, and also in considering the remedial action design to ensure that activities do not adversely affect these organisms.

2.5 Floral Surveys.

Plants present in the vicinity of a hazardous waste site can be identified and if desirable, counted. This information can be used to aid in delineating the type of habitat present. As for animals, particular attention is generally given to plants on State or Federal Threatened and Endangered Species lists. Areas on site where stressed vegetation is observed are given close attention and the source of the stressed vegetation should be determined. In many cases, sandy or nutrient-poor soils used as fill may be the cause of stressed vegetation but chemicals such as the petroleum hydrocarbons, certain metals, and herbicides can also impact vegetation.

2.6 Surveys of Local Surface Water Bodies.

Bodies of water can be surveyed for the presence or absence of key indicator species (for example, as noted above for the RBP) and for such factors as species relative abundance and species diversity. Site-specific values can be compared with either literature values or with data obtained from a reference water body to determine if any impairment has occurred. Organisms collected from water bodies can also be assessed as to their body burden of target chemicals (primarily chemicals that bioaccumulate). For example, the absence of large, top predatory fish and the presence of elevated levels of bioaccumulating chemicals in smaller fish would raise concerns.

3.0 ADVERSE EFFECTS

An important point to remember in ecological risk assessment is that there are numerous sources of perturbations in the environment (141). Toxic chemicals are just one of the many factors that can affect an ecosystem. For example, stressed vegetation at a facility may be viewed as an indication of chemical stress when it actually indicates low nutrient levels because the site was constructed on subsurface soils, either fill material or in an excavated area, and these soils have lower nutrient levels and water holding capacity. Another point to remember is that national toxicity criteria may not be appropriate for the particular piece of the nation in which a site is located and that organisms are adaptable (140). Many of the streams in Colorado downgradient from mining sites contain heavy metals at levels well above national water-quality criteria but apparently nobody has bothered to tell the fish that they should be dead.

Agents responsible for adverse effects can include such things as site chemicals, fill material used at the site, construction activities, and changes in physical properties of the area caused by the site (e.g., a change in pH or temperature). Ecological risk assessments tend to focus on chemical impacts but other sources of perturbations to the system can be equally important. For example, the effects of remedial activities on the site and surrounding areas needs to be evaluated prior to implementation. Excavating contaminated soils from a marshy area to protect human health under worst-case and highly unlikely potential future use scenarios, and destroying the marsh in the process, certainly do not achieve the intent of most cleanup regulations.

Toxic chemicals are of major concern as environmental contaminants if they either cause a potentially irreversible decline in a species or several species, or if they cause a decline in the aesthetics of an area. Toxicants can adversely affect a species by causing poisoning or by effecting a decline in the food supply for the species. Poisoning can be either by direct exposure in a contaminated area or, for predators or scavengers, by secondary exposure to contaminated prey or carrion. The latter method of exposure is of major concern because many persistent chemicals can be biomagnified to toxic levels and because top predators and scavengers are fewer in number than prey species and, in addition, may be less able to adjust to a declining population by increasing reproductive output. The classic example of this type of effect is the decline in predatory birds in the United States and Europe due to ingestion of prey contaminated with the persistent organochlorine pesticides (142,143). Another less visible but probably equally important example is the decline in predatory wasps that occurs following crop spraying. This decline in predators probably facilitates the recovery of the crop pests and then requires respraying of the field. The economic losses associated with this continuing scenario are probably dramatic. In addition to species at the top of the food chain, toxic effects on any endangered or threatened species are of major concern because even a small decline in the population of these species may have a drastic adverse effect.

A major problem in studying the effects of toxicants on wildlife is that it is often difficult to document sublethal and even lethal effects on wildlife in the field. In addition, even if an effect is seen, it can be very difficult to determine its cause. However, the results of laboratory studies do allow the development of some hypotheses about the potential effects of particular chemicals on wildlife (136,140,144). Toxicity tests conducted on laboratory animals and designed to evaluate potential effects on humans can be extrapolated to wildlife at least to some degree. Therefore, health effects information can be used to determine potential toxic effects in terrestrial wildlife. Lethality is the primary effect evaluated in aquatic organisms, and generally involves determining an LC50 for a 48- or 96-hour exposure. Although lethality is obviously important, in the natural aquatic environment subtle changes caused by sublethal effects may be more common.

4.0 PERSPECTIVE

Ecological risk assessment is becoming increasingly important as a part of hazardous waste site evaluations. In general, this increased emphasis is valuable to ensure that ecosystems are adequately protected. However, there are currently no detailed guidelines on such things as how assessments are to be conducted, the level of detail required, and what endpoints should be evaluated. In addition, a huge variety of organisms can be considered and numerous conditions (including but certainly not limited to chemicals) can affect these organisms. The level of uncertainty in ecological risk assessment may well dwarf the uncertainty in human health risk assessment. Consequently, it may never be possible to definitively determine whether or not a particular chemical will cause a particular effect at a site.

Having noted that it may never be possible to determine with certainty the effects of a chemical at a site, it is equally important to note that a conclusion of "further study is needed to determine if..." provides little useful information to a risk manager. The ecological risk assessor needs to ensure that the assessment is focused on solving rather than just studying a situation. Development of clear objectives, the use of common sense, and a consideration of all available information are critical to arriving at an appropriate decision for a site.

5.0 GENERAL REFERENCES

Environmental Protection Agency (EPA), *Risk Assessment Guidance for Superfund: Environmental Evaluation Manual, Interim Final.* Office of Emergency and Remedial Response. EPA 540/1-89/001A (1989).

Environmental Protection Agency (EPA), *Framework for Ecological Risk Assessment.* Risk Assessment Forum, USEPA, Washington, D.C. PA 630/R-92/001, February (1992).

Warren-Hicks, W., Parkhurst, B.R., and Baker, S.S., eds. 1989. *Ecological Assessments of Hazardous Waste Sites: A Field and Laboratory Reference Document.* Prepared by Kilkelly Environmental Associates, Raleigh, N.C. for the Environmental Protection Agency Office of Research and Development. EPA 600/3-89/013 (1989)

APPENDIX A - CASE STUDY 1

Baseline Site-Specific Risk Assessment for a Crop Duster Airfield

This case study presents a baseline health risk assessment (HRA) designed to address both current and potential future risks associated with a site under several possible scenarios. The sample HRA illustrates the procedures outlined in Chapters 1 - 5 for conducting site-specific risk assessments at sites contaminated with hazardous wastes, in this case, pesticides.

1.0 INTRODUCTION

The purpose of this baseline health risk assessment (HRA) is to evaluate the potential health risks to area residents posed by the presence of chemicals in soils at the Crop Duster Airfield site. This assessment evaluates the potential for exposure and risk under several scenarios for potential site use. These are:

<u>Current Conditions</u>

1. Short-term exposure - An analysis of health risks associated with short-term exposure to the site as it currently exists. (This scenario was performed as a preliminary risk assessment to address potential risks during an interim period prior to remediation).

2. Longer-term exposure - An analysis of health risks posed by the site if no remedial action is taken (a baseline risk assessment of current conditions).

Potential Future Conditions

3. Commercial development - An analysis of the potential health risks associated with development of the property for commercial office use. This scenario assumes that hazardous material would be capped on site under an asphalt parking structure or a building.

4. Residential use - An analysis of health risks associated with construction of residential dwellings on the site (baseline risk assessment - potential future use conditions).

Preparation of a risk assessment consists of combining site-specific information with standard risk assessment assumptions and formulas to estimate potential exposure, and then combining the exposure estimate with toxicity information to estimate risks. A key part of this risk assessment process is an evaluation of the uncertainties associated with the assessment process. This risk assessment was conducted in compliance with standard risk assessment documents, including:

Environmental Protection Agency (EPA) 1988. Superfund Exposure Assessment Manual. Office of Remedial Response, USEPA, Washington, D.C. April 1988. EPA 540/1-88-001

Environmental Protection Agency (USEPA) 1989. Exposure Factors Handbook. Office of Health and Environmental Assessment, Washington, D.C. July 1989. EPA 600/8-89/043

Environmental Protection Agency (USEPA) 1989. Risk Assessment Guidance for Superfund: Human Health Evaluation Manual, Part A (Interim Final). Office of Solid Waste and Emergency Response, Washington, D.C. July 1989. OSWER Directive 9285.701A

Environmental Protection Agency (USEPA) 1991. Human Health Evaluation Manual, Supplemental Guidance: "Standard Default Exposure Factors" Office of Solid Waste and Emergency Response, Washington, D.C. March 1991. OSWER Directive 9285. 6-03

Other EPA guidance documents and risk assessment literature were used if particular parameters have not been estimated by EPA or if other values appeared warranted.

Section 2 of the risk assessment contains background information on the site and information on the nature and extent of contamination. In Section 3, the environmental fate and transport of the key site contaminants and the key toxicological characteristics of these compounds are discussed. As part of the toxicological assessment section, health-based criteria, and other established criteria and guidance are identified. An exposure assessment, including a discussion of exposure pathways, a description of potential receptors, and a quantitative estimate of exposure, are presented in Section 4. Section 5 contains the risk characterization, in which information on the toxicity of pesticides of concern (from Section 3) is combined with the results of the exposure assessment (from Section 4) to determine if the site currently poses a risk to public health. A discussion of the uncertainties in the risk assessment is also included in this section. The conclusions of the risk assessment are presented in Section 6.

2.0 SITE CHARACTERIZATION

2.1 Site Background

The site consists of a former crop duster airport. The site is located on 42 acres of land, of which approximately 4 acres were used for operations. The airport is located in a mixed residential, industrial, and agricultural area. The airport was used for crop dusting operations from 1945 until 1980 by a number of crop dusting operators. Operations at the airfield consisted of storage of agricultural chemical products, loading of the chemicals into the planes, and routine airplane maintenance.

After spraying agricultural chemicals, airplanes were pressure-washed with soap and water on a wash pad located adjacent to the former maintenance building. Rinseate from the wash pad drained through a sump to a rinseate water-evaporation pond. Use of the rinseate pond ceased in 1984, and in 1988, pesticide-containing soils were excavated from the pond area and sent to a hazardous waste landfill. More recently, the airfield and associated buildings have been demolished and the site is currently a large vacant lot.

The 42 acres that comprise the site are located on an approximately 130-acre parcel of land that is bounded by roads (Figure 1). An orchard is located across the road to the northwest and east of this orchard is open land that is currently being considered for development for residential use. A middle school, a Little League baseball field and a truck maintenance shed are located east of the property. A town park is located south of the airfield. Ten homes are located across the road to the west of the site.

2.2 Hydrogeology

Surface vadose (unsaturated) zone soils consist of fine to medium grained silty sands and sands to a depth of 10 feet below ground surface. Increased amounts of silt and clay are present in soils below 10 feet and a moderately stiff clayey silt layer was reported at 17 feet. Groundwater is present at a depth of approximately 60 feet beneath the site. Evaluations in the area of the former rinsewater sump indicate that pesticides were generally not detected below 8 feet.

2.3 Nature and Extent of Pesticide Residuals

Several studies have been conducted to evaluate the nature and extent of any chemical residuals at the site. The most recent sampling was used as the basis for this assessment. Much of the other data on the site was collected as part of the previous remediation effort and is focussed on the area of the former evaporation pond.

Local consulting engineers collected samples primarily from areas around the former operations facility but also collected several samples at points along the runway and at the northwest corner of the airfield. Samples were analyzed for the presence of organochlorine pesticides by EPA Method 8080. Analysis indicated that only toxaphene and DDT and its metabolites DDE and DDD (ΣDDT) were widespread at the site. Table 2-1 summarizes ΣDDT and toxaphene results for samples from different depths at the airport site.

Analysis of these data suggests that the majority of the DDT and toxaphene is present in the upper 6 inches of soil and only rarely are pesticides detected below 12 inches. For DDT, 81/97 samples contained detectable levels of DDT at 0-6 inches, 55/97 sampling locations were positive at 6-12 inches, 11/97 at 12-24 inches and only 1/97 at 24-36 inches. It should be noted that not all samples collected from deeper soils were analyzed. Generally, samples were not analyzed if soils immediately above the sample did not contain detectable levels of pesticide or if pesticide concentrations were very low (generally less than 1 mg/kg for DDT, always non-detectable for toxaphene). Review of the data set also indicates that DDT is fairly widely spread throughout the site although high concentrations tend to be located near the former operations area. Toxaphene tends to be concentrated in an area immediately south of the former location of the paved taxi area and the closed mix system area and is generally detected only sporadically in other areas of the site.

TABLE 2-1
DDT and Toxaphene Concentrations in Soils at the Cropduster Airport

0-6 Inch Samples				
DDT	Mean: 9.5 mg/kg		GeoMean:	0.9mg/kg
	Std Err:	2.9	Count:	81/97
	95% UCL:	15 mg/kg	Maximum:	233 mg/kg
Toxaphene	Mean:	47 mg/kg	Geo Mean:	0.5 mg/kg
	Std Err:	17	Count:	19/97
	95% UCL:	84 mg/kg	Maximum:	1100mg/kg
6-12 Inch Samples				
DDT	Mean:	0.7 mg/kg	Geo Mean:	0.1 mg/kg
	Std Err:	0.3	Count:	55/97
	95% UCL:	1.3 mg/kg	Maximum:	31 mg/kg
Toxaphene	Mean:	8.3 mg/kg	Geo Mean:	0.3 mg/kg
	Std Err:	5.0	Count:	12/97
	95% UCL:	18 mg/kg	Maximum:	390 mg/kg
12-24 Inch Samples				
DDT	Mean:	0.6 mg/kg	Geo Mean:	0.1 mg/kg
	Std Err:	0.5	Count:	11/97
	95% UCL:	1.7 mg/kg	Maximum:	52 mg/kg

Toxaphene	Mean:	4.9 mg/kg	Geo Mean:	0.3 mg/kg
	Std Err:	3.8	Count:	6/97
	95% UCL:	12 mg/kg	Maximum:	364

The term ΣDDT is used to refer to the sum of DDT, DDD, and DDE. The detection limit was 0.4 mg/kg for ΣDDT and 0.4 mg/kg for toxaphene. A value of half the detection limit was used in the calculations for values reported as not detected.

3.0 CHARACTERISTICS OF KEY CHEMICALS

3.1 Environmental Behavior

The environmental behavior of chemicals is dependent on the physical and chemical properties of the compounds, the environmental transformation processes affecting them, and the characteristics of the media (soil, sediments, water) in which they are present and through which they may migrate. DDT and toxaphene initially applied to or spilled on soils will volatilize under climatic conditions like those present at the Airfield (EPA 1979; ATSDR 1989, 1990). The pesticide in this area has been exposed to high temperatures, sunlight, and rain, and has been present for a considerable amount of time. It seems likely that during that time any DDT or toxaphene that was not tightly bound to soils would have volatilized at least from the surface layer of soil. Organochlorine pesticides still remaining in the soils are likely to be rather tightly bound to soil particles. These compounds may not be as readily bioavailable as freshly spilled pesticides.

The principal environmental fate processes for the organochlorine pesticides are adsorption to soils, runoff into any nearby streams and deposition in sediments, or migration and redeposition as a component of windborne dust. The organochlorine pesticides are not considered to be very mobile in the environment and are likely to remain bound to site soils.

3.2 Background Levels in Farmland Soils

Organochlorine pesticides were widely used for pest control between 1945 and approximately 1970. These materials are highly persistent in the environment and they are still encountered regularly in soils in regions where these chemicals were sprayed to control insect pests such as crop pests, mosquitoes and gypsy moths (Carey et al 1979). A survey of pesticide residues in croplands, conducted in 1972, showed that DDT/DDD/DDE were present in 70% of all samples at an average concentration of 0.49 mg/kg. Toxaphene was detected in 14% of the samples at an average level of 0.25 mg/kg.

3.3 Toxicity/Hazard Evaluation

Preparation of a toxicity/hazard assessment involves considering the types of adverse health effects associated with the chemicals of concern and identifying the amount (dose) of these chemicals required to produce these adverse effects. A brief description of the toxicology of ΣDDT and toxaphene focussing on their toxicity criteria is presented below. Toxicity criteria that have been validated by EPA were available for ΣDDT and for toxaphene, and these values were used in the assessment.

DDT and its Metabolites

p,p'-Dichlorodiphenyl trichloroethane (DDT), p,p'-Dichlorodiphenyl dichloroethane (DDD), p,p'-Dichlorodiphenyl dichloroethylene (DDE) are all classified as Group B2, probable human carcinogens, based on sufficient evidence of carcinogenicity in animals but inadequate evidence in humans. The EPA (1993) has calculated cancer slope factors for the compounds of 0.34 $(mg/kg/day)^{-1}$ for DDT and DDE and 0.24 $(mg/kg/day)^{-1}$ for DDD. Because the difference between 0.24 and 0.34 is well within the range of uncertainty in cancer slope factors, the DDT cancer slope factor of 0.34 $(mg/kg/day)^{-1}$ will be used for all DDT-related compounds. The EPA (1993) calculated a reference dose (RfD; formerly referred to as an ADI) for DDT of 5×10^{-4} mg/kg/day but did not develop RfDs for the other DDT-related compounds. The RfD for DDT will be used for ΣDDT.

Toxaphene

The EPA (1993) classified toxaphene in Group B2, probable human carcinogen, based on sufficient evidence of carcinogenicity in animals but inadequate evidence in humans. EPA has calculated a cancer slope factor for toxaphene of 1.1 $(mg/kg/day)^{-1}$ based on the production of liver tumors in mice. The State Environmental Regulatory Agency has determined a cancer slope factor for toxaphene of 1.2 $(mg/kg/day)^{-1}$ based on the same study. These values are essentially equal but for political reasons (the site is a State-lead site) the state value will be used in this risk assessment. Neither the State nor EPA have determined an RfD for toxaphene.

Toxicity criteria are based on a large number of assumptions and it is important to note the potential effects of these assumptions on the toxicity values. For example, the current approach to assessing the potential risk associated with exposure to carcinogens that is employed by EPA and most regulatory agencies, is to assume that no threshold dose (a dose below which no effects will occur) exists. This approach is based on the theory that carcinogens act by mutating cells and that even a single molecule of a carcinogen can cause a mutation and therefore poses some risk. It should be noted however that not all scientists agree that a non-threshold approach is correct for all carcinogens. For example, some researchers feel that for chemicals that do not appear to be mutagenic (such as most organochlorine pesticides including DDT and possibly toxaphene), the increased cancer incidences observed in animal bioassays may simply be a result of increased cell proliferation caused by the toxic effects of the high doses of chemicals administered to the animals. If this theory is accurate, a threshold level would exist below which exposure would not cause cancer. It is currently not possible to determine which of these theories is correct. Therefore, to ensure that public health is protected, the EPA and most states use a nonthreshold approach to risk assessment for all carcinogens. However, it is important to note that the difference between these two theories introduces a considerable amount of uncertainty, in the form of a health-protective bias, into the risk assessment.

4.0 EXPOSURE ASSESSMENT

The objectives of the exposure assessment are to identify actual or potential exposure pathways, to characterize the potentially exposed populations, and to determine the extent of any exposure. An exposure route consists of four main elements:

(1) a source and mechanism of chemical release to the environment;
(2) an environmental transport medium for the released chemical;
(3) a point of actual or potential contact by a receptor with the contaminated medium (referred to as the exposure point); and
(4) a method of intake (e.g., ingestion, inhalation, dermal absorption) at the exposure point.

If any one of these factors is absent, exposure cannot occur because the receptor will not be able to come in contact with the chemical. The potential for chemicals to get from the site soils to a receptor population is evaluated as the first section of an exposure assessment.

Potential receptor populations for the site include all people in the area of the site who may be exposed to chemicals either by direct contact on site or as a result of exposure to chemicals emanating from the site. However, the level of exposure experienced by different individuals is likely to vary considerably. It is standard practice in site-specific risk assessments to evaluate a reasonable maximum exposure to an individual rather than an average exposure to a population. Use of the reasonable maximum individual exposure ensures that most, if not all people in the area of a site receive actual exposures that are equal to or lower than the estimated reasonable maximum exposure levels. In many cases, actual exposure may be zero.

Once the exposure routes and potential receptors have been identified, the final step in an exposure assessment is a determination of the potential levels of exposure that are associated with the various possible exposure routes. Exposure quantification is presented in Section 4.2.

4.1 Exposure Routes

As noted above, the initial step in an exposure assessment is a determination of all routes that may lead to exposure at the site followed by an evaluation of the likelihood of exposure occurring via any of these routes. For the Crop Duster site, four general exposure scenarios are considered. These scenarios are:

Current Conditions

1. Short-term exposure - Exposure for a one year time period assuming no change in use patterns at the site.
2. Longer-term exposure - Exposure for a lifetime assuming no change in use patterns at the site.

Potential Future Conditions

3. Commercial development - Exposure that would be associated with development of the property for commercial office use. This scenario assumes that hazardous material would be capped on site under an asphalt parking structure or a building.
4. Residential use - Exposure that would be associated with construction of residential dwellings on the site (baseline risk assessment - potential future use conditions).

Possible routes of exposure at the field include direct contact with pesticide residuals in soil and inhalation exposure to chemicals in air for each of these scenarios. However, the relative amount of material contacted or inhaled is likely to vary considerably among the scenarios.

Direct contact with soils may lead to exposure via dermal absorption of chemicals present in the soils or via ingestion (either inadvertent or intentional) of on-site soils. Exposure to airborne chemicals could include inhalation of pesticide-containing windborne particulate matter. Ingestion of vegetables grown on site could occur if residents on site planted home vegetable gardens.

A route of exposure that is often considered at hazardous waste sites is the use of on-site groundwater as a source of potable water. This route is considered unlikely to occur at the property and is not evaluated in detail in the assessment. ΣDDT and toxaphene are relatively immobile in the environment and sampling indicates that the pesticides are generally present only in the upper 1 foot of soils. Groundwater at the site is encountered at approximately 60 feet below the surface and this water appears to meet State criteria for water quality. ΣDDT and toxaphene were not detected in groundwater monitor wells installed at the site.

Direct Contact - Surface Soils

Direct contact with surface soils can lead to exposure to persistent compounds such as the organochlorine pesticides through inadvertent or intentional soil ingestion or through dermal absorption. The potential ingestion of contaminated soils will be evaluated quantitatively in this assessment for each of the potential scenarios. Several authors have indicated that relative to ingestion, exposure via dermal absorption is minimal, particularly for compounds that are tightly bound to soil (LaGoy and Hopkins 1991; Sedman 1989). Consequently, the potential exposure via dermal absorption was not addressed in this assessment.

Air

Individuals on site may be exposed to DDT or toxaphene that volatilizes from site cover soils or that is present in any dusts that are released into the air as a result of wind erosion of surface soils. Because of the rather low concentrations of the chemicals and their chemical characteristics, volatilization is considered unlikely to be a significant exposure pathway at the airport site. The potential for exposure via inhalation of organochlorine pesticides that are components of windborne dust will be evaluated in this exposure assessment. The population of concern for exposure via this pathway depends somewhat on the particular exposure scenario being considered but generally includes residents living in the vicinity of the site in a downwind direction and on-site workers or residents.

Vegetable Consumption

If vegetables are grown on the site under a residential use scenario, the organochlorine pesticides in the soils may be translocated to the vegetables and subsequently ingested by people consuming these vegetables. Under certain conditions, residents may receive their highest potential exposure by consuming vegetables from a home garden. However, this pathway was not evaluated in detail for this assessment. A large amount of uncertainty is associated with the pathway, and because of the large uncertainty, the pathway would contribute disproportionately to risks at the site. Uncertainty is associated with differences in uptake for different vegetables, variability in gardening habits and vegetable consumption habits and with factors such as bioavailability of chemicals. The fact that this pathway was not included in the assessment is discussed as an uncertainty in Section 5.2.

4.2 Quantitative Exposure Assessment

Individuals are exposed to materials released into the environment in varying quantities and proportions, via a wide variety of exposure routes. The actual amount of material an individual is exposed to depends on the individual's frequency, extent, and duration of exposure which in turn depends on many factors, including location of residence, age, body weight, sex, and activity patterns. Patterns of exposure are highly variable among individuals. Standard approaches and assumptions were used in preparing the exposure assessment. These approaches and assumptions are designed to be health protective, that is, they are designed to produce estimates of exposure that overestimate rather than underestimate exposure and risk. The rationale used in selecting values for this assessment is discussed in each subsection. Table 4-1 presents a list of the parameters for assessing exposure that are used in the risk assessment.

A general principle of exposure assessment is that conservative assumptions (i.e., assumptions that tend to overestimate exposure and therefore risk) are made at each step, so that the final estimates of exposure will be higher than (often much higher than) the upper end of the likely range of actual exposures. Estimates of exposure developed in this chapter are conservative, in this sense, in two independent ways. First, for each exposure scenario, the assessment is focused on the locations where exposure is expected to be greatest, and on the subpopulations likely to be most exposed. For this reason, the estimates of exposure are really only valid for the most highly exposed persons, referred to as the maximally exposed individual (MEI) and will overestimate exposure for most people. Second, exposure assessments for these persons incorporate health protective assumptions about many factors that influence exposure, including emission rates, deposition rates, environmental persistence, bio-uptake, bioavailability, and human activity. For this reason, the exposure estimates are expected to provide a maximum estimate of exposure even for the MEI.

Exposure for each of the four possible exposure scenarios (two scenarios addressing exposure under current site conditions and three addressing possible future use of the site) is discussed in this section. Quantitative estimates of exposure are provided for all scenarios except use of the site for commercial office buildings (Scenario 3). A qualitative discussion of exposure is provided for each scenario.

Scenario 1: Current Conditions, Short-term exposure

Soil Ingestion

Exposure via direct ingestion of soils is of greatest concern if the soils are in an area that is frequently used (as in a backyard) and if the soils are not covered by grass, pavement, or buildings. On-site workers engaged in digging, gardening, or earth moving activities and local children playing on the site are most likely to contact soils. The site primarily consists of a flat area of uncovered soils. There do not appear to be any features on site that would be particularly appealing to children. In addition, use of the site by workers should be limited. Therefore, although the lack of vegetative cover will increase the potential for contact with soils, the frequency of exposure to site soils should be limited.

In order to estimate likely maximum exposure levels for children, it is assumed that a child is exposed to site soils 100 days/year (equal to both days of every weekend or every day of the summer) for a year. The time period of a year is based on the time from demolition of the airport to the time when remediation activities at the site are expected to begin. It is further assumed that children likely to enter the site are 5 years old (old enough to be alone but still ingesting soil at a higher rate than older children) and ingest 200 mg/day of soil. It is assumed that on the days they visit the site, all soil ingested comes from the site. Finally, it is assumed that the child weighs 15 kg during the period of exposure. Because of the assumptions used and differences in behavior between young children and older children, it is expected that lower exposure would be experienced by school-age children crossing the site on their way to and from school.

For workers, it is assumed that a worker is exposed 100 days/year to site soils for the 1 year period, and ingests 50 mg/day of soil (EPA 1991). It is further assumed that the worker weighs 70 kg and has an average lifespan of 70 years (EPA 1989).

For noncarcinogens, exposure is averaged over the duration of exposure. For carcinogens, cancer potency factors are developed based on a full lifetime exposure and therefore the averaging time is also expressed in terms of a full lifetime of 70 years (EPA 1989, 1991). Using these assumptions, exposure can be estimated using the formula:

$$CDI = \frac{CS \times IR \times CF \times FI \times EF \times ED}{BW \times AT}$$

Where,

CDI = Average daily intake of the chemical (mg/kg/day)
CS = Chemical concentration in soil (mg/kg)
IR = Soil ingestion rate (200 or 50 mg/day)
CF = Conversion factor (10^{-6}kg/mg)
FI = Relative fraction absorbed from soil (100%)
EF = Frequency of site contact (100 days/365 days)
ED = Exposure duration (1 year)
BW = Body weight (15 or 70 kg)
AT = Averaging time (1 or 70 years)

Using this equation and the 95% upper confidence limit on the average soil concentration for the upper 6 inches of site soils from Table 2-1, chronic daily intake levels can be calculated and are presented in Table 4-2.

Inhalation Exposure

Exposure to site-related chemicals may occur as a result of inhalation of fugitive dust containing the chemicals. No ambient air monitoring has been conducted at the facility and therefore, simple air modelling default values were used to estimate air concentrations. Exposure was calculated for a house located on the edge of the site assuming 1 year of exposure and a constant migration of windborne dust from the site to the house.

Inhalation exposure will depend on the concentration of pesticides present in the air that is inhaled, and the amount of time an individual is exposed. No information is available on the amount of particulate matter actually airborne at the site. EPA (1991) has estimated maximum likely concentrations of airborne, site-derived particulate matter of 2.2×10^{-10} kg/m^3 using models for fugitive dust dispersion. Another approach that has been used is to assume that off-site air contains site-derived dust at the National Air Quality Standard level of 50 ug/m^3 (5×10^{-8} kg/m^3). For this assessment, we used the more conservative approach[1] and assumed that air contains site-derived dust at 50 ug/m^3. The average person weighs 70 kg and inhales 20 m^3 of air each day and may be exposed for 350 days each year (EPA 1991). A value of 1 year was used for the duration of exposure. For non-carcinogenic effects, exposure is averaged over the duration of exposure while for carcinogens, the averaging time is expressed in terms of a full lifetime of 70 years (EPA 1989).

Using these assumptions, exposure can be estimated using the formula:

---

[1]ed. note: An initial evaluation using EPA's fugitive dust dispersion value showed risks in the range of 10^{-10}, far below any regulatory risk goals. However, the State reviewer was uncertain of the derivation of the EPA value and to prevent further discussion of a clearly insignificant risk, the more conservative approach was used.

$$CDI = \frac{SC \times PA \times BR \times EF \times ED}{BW \times AT}$$

where,

CDI = Inhalation dose (mg/kg/day)
SC = Concentration of chemical in soil (mg/kg)
PA = Concentration of dust in air (5×10^{-8} kg/m^3)
BR = Breathing rate (20 m^3/day)
EF = Exposure frequency (350/365 days)
ED = Exposure duration (1 year)
BW = Body weight (70 kg) and
AT = Averaging time (1 or 70 years)

The soil concentrations used in the equation were the upper 95% confidence limit of the average soil values for the upper soils from the site. These values are presented in Table 2-1. The inhalation doses calculated using this equation are presented in Table 4-3.

Scenario 2: Current Conditions, Long-term exposure

Soil Ingestion

As noted above, the frequency of exposure to soils at the site should be limited under current conditions. In order to estimate likely maximum exposure levels for children and adult workers over longer time periods, it is assumed that a child is exposed to site soils 100 days/year (equal to both days of every weekend or every day of the summer) for 15 years. The time period of 15 years is based on the time from when children are old enough to enter the site (5 years old) to adulthood (approximately age 20). People of this age ingest 100 mg/day of soil (EPA 1991). It is assumed that on the days they visit the site, all soil ingested comes from the site. Finally, it is assumed that the child weighs an average of 40 kg over the entire period of exposure.

For workers, it is assumed that a worker is exposed 50 days/year to site soils for a 10 year period, and ingests 50 mg/day of soil (EPA 1991). It is further assumed that the worker weighs 70 kg and has an average lifespan of 70 years (EPA 1989).

Inhalation Exposure

Inhalation exposure was calculated for a house located on the edge of the site assuming 30 years of exposure (EPA's (1991) assumption for maximum likely residence in one house) and a constant migration of windborne dust from the site to the house. As for short-term exposure, actual inhalation exposure will depend on the amount of pesticides present in the air that is inhaled, and the amount of time the individual is exposed. The same assumptions used to calculate short-term exposure can be used to estimate inhalation exposure.

Using these assumptions, exposure can be estimated using the formulas presented in Scenario 1. Using these equations and the 95% upper confidence limit on the average soil concentration for the upper 6 inches of site soils from Table 2-1, chronic daily intake levels can be calculated for soil ingestion and inhalation and are presented in Tables 4-4 and 4-5.

Scenario 3: Potential Future Use, Commercial Offices

One possible use for the former airfield is the construction of office buildings on the property. The airfield is located in an area of mixed residential and commercial office use. If the site was to be used for this purpose, some limited exposure to workers could occur during construction activities. However, this exposure is likely to be of short duration and frequency and is not expected to pose substantial risks based on review of the other scenarios. Once the buildings are constructed, it is assumed that the bulk of the site would be covered by buildings or asphalt parking lots and the remaining area would be landscaped. If the construction is designed to ensure capping and containment of the pesticide-containing soils, the potential for exposure is essentially non-existent. Even if construction activities are not focused on providing pesticide containment, most site soils will be covered and it is unlikely that individuals would come in contact with the soils on a repeated or frequent basis. Consequently, exposure under this scenario appears to be unlikely and if it occurs at all, almost certainly poses risks well below the State and EPA target risk level of 10^{-6}.

Scenario 4: Potential Future Use: Residential exposure

Soil Ingestion

Exposure via direct ingestion of soils is of greatest concern if the soils are in an area that is frequently used (as in a backyard). Consequently, the residential use scenario is generally considered likely to pose the greatest risk. Exposure can occur to very young children who ingest substantially higher amounts of soil than older children. In addition, landscaping or gardening may also lead to higher exposure than would occur in undisturbed areas. Because of this increased potential for exposure, noncarcinogenic risks will be evaluated based on exposure to children. Under the residential use scenario, it is assumed that a 1-6 year old child is exposed to soils 350 days/year for five years. It is further assumed that children of this age ingest 200 mg/day of soil. It is assumed that all soil ingested comes from the former site. Finally, it is assumed that the child weighs 15 kg during the period of exposure.

For carcinogenic effects where exposure is averaged over a lifetime rather than over the exposure period, both childhood and adult exposure is considered. The standard approach used by the USEPA is to assume that an individual resides in the same location for 30 years, a time period that represents the upper 90th% value for time in a single residence. Over this time period it is assumed that an individual ingests soil at 200 mg/day and weighs an average of 15 kg for the 5 years between the ages of 1-6, ingests 100 mg/day of soil and weighs 30 kg between the ages of 6 and 11, and ingests 100 mg/day of soil and weighs 70 kg between the ages of 11 and 30. It is further assumed that the individual is exposed 350 days/year to site soils and has an average lifespan of 70 years (EPA 1989).

Using these assumptions, exposure can be estimated using the formulas presented in Scenario 1, modified to account for variability in intake rates and body weight with age:

$$\mathrm{CDI} = \frac{\mathrm{CS \times CF \times FI \times EF}}{\mathrm{AT}} \times \sum \frac{\mathrm{IR_i \times ED_i}}{\mathrm{BW_i}}$$

Where,

CDI, CS, CF, FI, EF, and AT are as described in Scenario 1 and IR_i,

ED_i, and BW_i are the ingestion rate, exposure duration, and body weight for the ith exposure period.

Using these equations and the 95% upper confidence limit on the average soil concentration for the upper 6 inches of site soils from Table 2-1, chronic daily intake levels can be calculated for soil ingestion and inhalation and are presented in Tables 4-6 and 4-7.

TABLE 4-1

Exposure Assumptions for Use at the Crop Duster Site

Parameter	Exposure Assumption[a]
Scenario 1	
Soil Ingestion	
Frequency of Exposure	100 visits/365 days (1 year)
Exposure Duration	1 year
Age of Child	5 years
Average Weight over Period of Exposure	15 kg (Child)[b] 70 kg (Adult)
Incidental Ingestion of Soil	200 mg/day (Child)[c] 50 mg/day (Adult)
Percent of Contaminant Absorbed from Soil Ingestion	100%
Lifetime	70 years[c]
Inhalation	
Breathing Rate	20 m^3/day

TABLE 4-1 - (Scenario 1, continued)

Parameter	Exposure Assumption[a]
Concentration of Dust from Site that Becomes Airborne	5 x 10^{-8} kg/m^3
Exposure Frequency	350 days/365 days
Exposure Duration	1 year
Body Weight	70 kg

Parameter	Exposure Assumption[a]
Scenario 2	
Soil Ingestion	
Frequency of Exposure	
Children - Age 20	100 visits/365 days for 15 years
Adults	50 visits/365 days for 10 years
Age of Child	5-11 years (systemic effects) 5-20 years (carc. effects)
Age of Adult	20+ years (adults)
Average Weight over Period of Exposure	25 kg (5-11 year child)[b] 40 kg (5-20 year youth) 70 kg (Adult)
Incidental Ingestion of Soil	100 mg/day (Age 6-11)[c] 50 mg/day (Adult)
Percent of Contaminant Absorbed from Soil Ingestion	100%
Exposure Duration	30 years
Lifetime	70 years[c]

Table 4-1 - (Scenario 2, continued)

Parameter	Exposure Assumption[a]
<u>Inhalation</u>	
Breathing Rate	20 m^3/day
Concentration of Dust from Site that Becomes Airborne	5 x 10^{-8}kg/m^3
Exposure Frequency	350 days/365 days
Exposure Duration	30 years
Body Weight	70 kg

Parameter	Exposure Assumption[a]
Scenario 4	
Soil Ingestion	
Frequency of Exposure	350 days/365 days
Age	1-6 years (child) 20+ years (adults)
Average Weight over Period of Exposure	15 kg (1-6 year child)[b] 30 kg (5-20 year youth) 70 kg (Adult)
Incidental Ingestion of Soil	200 mg/day (Age 1-6)[c] 100 mg/day (Age 6+)
Percent of Contaminant Absorbed from Soil Ingestion	100%
Exposure Duration Lifetime	30 years 70 years[c]

TABLE 4-1- (Scenario 4, continued)

Parameter	Exposure Assumption[a]
Inhalation	
Breathing Rate	20 m^3/day
Concentration of Airborne Dust	5 x 10^{-8}kg/m^3
Exposure Frequency	350 days/365 days
Exposure Duration	30 years
Body Weight	70 kg

[a]These exposure assumptions are designed to represent reasonable maximum exposure conditions. If no risk is associated with exposure under these conditions, it is unlikely that exposure poses a risk at the site. If use of these values suggests that a risk may be present, a more careful analysis of the assumptions and uncertainties in the assessment should be conducted.

[b]Anderson et al. (1985); EPA (1989).

[c]EPA 1989; 1991

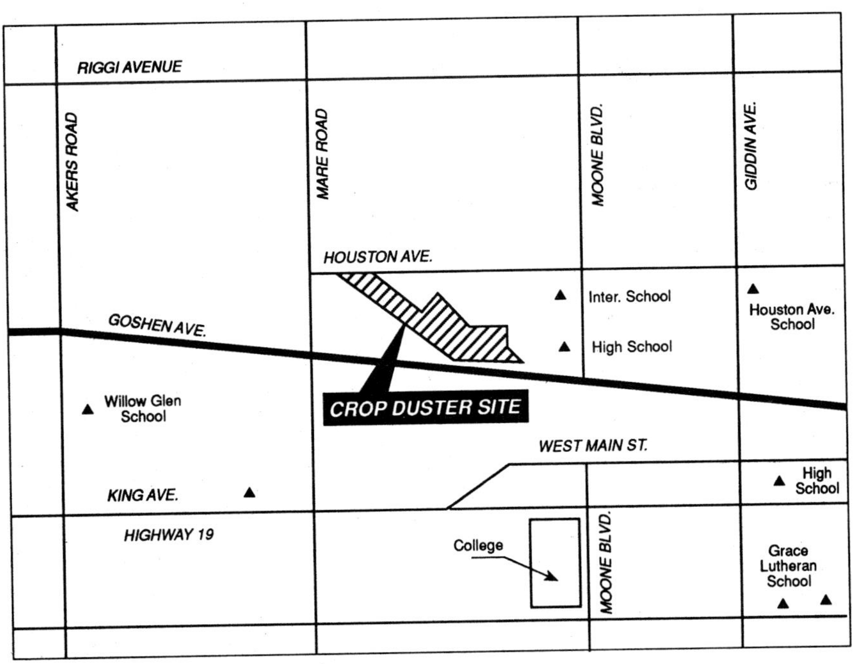

Figure 1
SITE LOCATION MAP
Crop Duster Airfield Site

TABLE 4-2

Exposure from Soil Ingestion at the Crop Duster Site

Scenario 1 - Short term Exposure, Current Conditions

Children

	Soil Conc.	Chronic Daily Intake
Non-carcinogenic exposure estimate		
DDT:	15 mg/kg	5.5×10^{-5}mg/kg/day
Carcinogenic exposure estimate		
DDT:	15 mg/kg	7.8×10^{-7}mg/kg/day
Toxaphene:	84 mg/kg	4.4×10^{-6}mg/kg/day

Adult Workers

	Soil Conc.	Chronic Daily Intake
Non-carcinogenic exposure estimate		
DDT:	15 mg/kg	3.0×10^{-6}mg/kg/day
Carcinogenic exposure estimate		
DDT:	15 mg/kg	4.3×10^{-8}mg/kg/day
Toxaphene:	84 mg/kg	2.3×10^{-7}mg/kg/day

TABLE 4-3

Exposure to Airborne DDT and Toxaphene at the Crop Duster Site

Scenario 1: Short-term Exposure, Current Conditions

Compound	Chronic Daily Intact
Non-carcinogenic exposure estimate	
DDT:	1.9×10^{-7}mg/kg/day
Carcinogenic exposure estimate	
DDT:	2.9×10^{-9}mg/kg/day
Toxaphene:	1.6×10^{-8}mg/kg/day

TABLE 4-4

Exposure from Soil Ingestion at the Crop Duster Site

Scenario 2 - Long-term Exposure, Current Conditions

Children

	Soil Conc.	Chronic Daily Intake
Non-carcinogenic exposure estimate		
DDT:	15 mg/kg	3.3×10^{-5}mg/kg/day
Carcinogenic exposure estimate		
DDT:	15 mg/kg	2.2×10^{-6}mg/kg/day
Toxaphene:	84 mg/kg	1.3×10^{-5}mg/kg/day

Adult Workers

	Soil Conc.	Chronic Daily Intake
Non-carcinogenic exposure estimate		
DDT:	15 mg/kg	1.5×10^{-6}mg/kg/day
Carcinogenic exposure estimate		
DDT:	15 mg/kg	4.3×10^{-7}mg/kg/day
Toxaphene:	84 mg/kg	2.3×10^{-6}mg/kg/day

TABLE 4-5

Exposure to Airborne DDT and Toxaphene at the Crop Duster Site

Scenario 2: Long-term Exposure, Current Conditions

Compound	Chronic Daily Intake
Non-carcinogenic exposure estimate	
DDT:	1.9×10^{-7}mg/kg/day
Carcinogenic exposure estimate	
DDT:	8.7×10^{-8}mg/kg/day
Toxaphene:	4.8×10^{-7}mg/kg/day

TABLE 4-6

Exposure from Soil Ingestion at the Crop Duster Site

Scenario 4 - Residential Exposure, Potential Future Use

Children

	Soil Conc.	Chronic Daily Intake
Non-carcinogenic exposure estimate		
DDT:	15 mg/kg	1.9×10^{-4}mg/kg/day
Resident Between the Ages of 1 - 30 Years		
Carcinogenic exposure estimate		
DDT:	15 mg/kg	2.7×10^{-5}mg/kg/day
Toxaphene:	84 mg/kg	1.4×10^{-4}mg/kg/day

TABLE 4-7

Exposure to Airborne DDT and Toxaphene at the Crop Duster Site

Scenario 4 - Residential Exposure, Potential Future Use

Compound	Chronic Daily Intake
Non-carcinogenic exposure estimate	
DDT:	1.9×10^{-7}mg/kg/day
Carcinogenic exposure estimate	
DDT:	8.7×10^{-8}mg/kg/day
Toxaphene:	4.8×10^{-7}mg/kg/day

5.0 RISK EVALUATION

In this section, information on the potential levels of exposure to contaminants (presented in Section 4) is combined with information on the toxicity of the contaminants (presented in Section 3) in order to determine the potential health risks to individuals living near the site under current conditions. For noncarcinogenic (systemic) effects, EPA assumes that there is a level below which no effects will occur (a threshold no-effect concentration). For carcinogens, however, it is assumed that any exposure has some probability of causing cancer and it is therefore assumed that there is no threshold level. Because of these different assumptions, potential noncarcinogenic and carcinogenic health effects are evaluated separately. The cancer risk is obtained by multiplying the chronic daily intake of the contaminant under consideration (from the exposure assessment) by its cancer potency factor or slope factor (from the toxicity section). EPA typically expresses cancer risk in terms of an upper-bound excess lifetime cancer risk level. EPA and the state use an upper-bound excess lifetime risk level of a one in one million (10^{-6}) as a risk level goal for carcinogens and EPA generally sets regulatory criteria within a risk range of between one in ten thousand and one in one million (10^{-4} and 10^{6}).

To evaluate possible risk from exposure to noncarcinogenic contaminants, the chronic daily intake (CDI; the average daily exposure dose) is divided by the health criterion value [the reference dose (RfD)]. If the CDI:RfD ratio (also called the hazard quotient or HQ) is less than one (i.e., if the daily intake is below the health criterion) the contaminant is considered unlikely to pose a health hazard to individuals exposed under the given scenario. As suggested in the EPA (1986) guidelines for Health Risk Assessment of Chemical Mixtures, the HQs are summed to determine if combined exposure may pose a health concern (USEPA 1986). It should be noted that the summed HQs (termed the hazard index or HI) is used only as an indication of possible hazard, as summation is only valid if there are no synergistic or antagonistic interactions among the summed compounds and if they have the same mechanism and site of action.

In addition to summing the risks posed by different chemicals, individuals may be exposed to more than one media by more than one route. For example, a resident could receive exposure to pesticides in soils via ingestion and could also be exposed to any airborne contaminants.

A number of assumptions were used in deriving both the CDI and the RfD or slope factor. Therefore, there is some uncertainty in interpreting the implications that the cancer risk levels and noncarcinogen HQs have for risk. However, in the estimation of these factors, conservative (health protective) assumptions were made so that it is improbable that the cancer risks would be higher or the HQs would be less than one if there were any actual hazard at the site. A discussion of the uncertainties in risk assessment is presented in Section 5.2.

5.1 Risks at the Site

Risks associated with exposure to ΣDDT and toxaphene under the various exposure scenarios are presented in Tables 5-1 through 5-6. As noted above, noncarcinogenic health effects and cancer risks are evaluated separately. For all scenarios, the HQs for the noncarcinogenic effects (the ratio of the exposure dose to a dose that poses a health concern) were less than one, indicating that systemic effects are unlikely. Similarly, the estimates of cancer risk for Scenario 1: Short-term exposure, and for inhalation exposure were all below the state and EPA target risk level of 10^{-6} excess cancer risk, suggesting that exposure under Scenario 1 and airborne exposure are unlikely to pose a significant risk at the site. For scenarios 2 and 4, however, the upperbound excess cancer risk for exposure via soil ingestion was above the 10^{-6} target risk level, although soil ingestion did not pose risks outside of the USEPA risk range for regulatory action (10^{-4} - 10^{-6}). No estimate of exposure and risk was developed for Scenario 3: potential future use as a commercial office building, because of the uncertainty in arriving at this value. However, a qualitative evaluation of the potential for exposure under this scenario suggests that risks above the 10^{-6} excess cancer risk level are unlikely. The majority of the site soils would almost certainly be covered by either a building or asphalt and consequently direct contact with soils would be impossible. A small amount of soil might be covered with only grass or landscaping but any contact with pesticide residuals in these areas would probably be of short duration and low frequency.

5.2 Uncertainties in the Risk Assessment

The procedures and inputs used to assess potential human health risks in this and most such evaluations are subject to a wide variety of uncertainties. In general, there are five main sources of uncertainty in the risk assessment of a reasonably well characterized site:

- -- Environmental chemistry sampling and analysis,
- -- Environmental parameter measurements,
- -- Fate and transport modeling,
- -- Toxicological data and dose-response extrapolations, and
- -- Errors through combinations of the above.

These sources of uncertainty as they pertain to this preliminary risk assessment are discussed below.

Environmental chemistry sampling and analysis error can stem from the error inherent in the procedures, from a failure to take an adequate number of samples to arrive at sufficient areal resolution, from mistakes on the part of the sampler, or from the heterogeneity of the matrix being sampled. Some small sampling and analytical error is almost certainly present, but overall, sampling and analysis error probably only increases the uncertainty in the assessment to a small degree.

Exposure estimation is another potentially large source of uncertainty in this risk assessment. The fact that consumption of vegetables from a home garden was not addressed may lead to underestimating exposure for individuals who consume large quantities of produce from an on-site garden. However, most of the uncertainty in the exposure calculations will lead to overestimating risk. The use of the 95% upper confidence limit of the average concentration of the pesticides at the site is likely to lead to overestimating the actual concentrations that individuals could contact. In addition, exposure estimates are highly dependent on the prediction of intake values, exposure frequency, exposure duration, and other exposure assumptions used in the assessment. The exposure parameters used in this exposure assessment, such as use of the site for 100 days/year, were selected to ensure that exposure was overestimated rather than underestimated. The extent of the overestimation will vary among individuals and is unknown, but may be substantial.

The results of animal studies were used to predict the potential health effects of a chemical in humans. Extrapolation of toxicological data from animal tests is probably one of the largest sources of uncertainty in any risk assessment. There may be important but unidentified differences in uptake, metabolism, and distribution in the body between the test species and man. Typically, the animals are administered high doses of a chemical in a standard diet while humans are generally exposed to much lower doses in a highly variable diet. Humans have a 70 year lifetime and may be exposed either intermittently or regularly for an exposure period ranging from months to a full lifetime. Because of these differences, it is not surprising that extrapolation error is a large source of uncertainty in risk assessment. Even if studies in humans are available, uncertainties can be large because the diet, activity patterns, exposure duration and frequency, and individual susceptibility may not be the same in the study populations as in the individuals exposed at the site.

Many carcinogenic compounds, including DDT, do not appear to be genotoxic. Toxaphene appears to be genotoxic, but some of the observed tumors, including the tumors used as the basis for risk calculations, appear to act via an epigenetic mechanism. EPA currently assumes that exposure to even very small amounts of any carcinogen poses some probability of causing cancer. Therefore, EPA's guidance for carcinogens assumes that no threshold (dose below which no effect would occur) exists for carcinogens. There is currently considerable debate on this issue and if an actual threshold does exist for certain compounds, the use of a nonthreshold approach in this assessment may substantially overestimate risk.

Uncertainties from different sources will be compounded in the risk assessment. For example, if a CDI for a contaminant measured in the environment is compared to an RfD to determine potential health hazard, the uncertainties in the concentration measurement, exposure assumptions and the toxicology are all expressed in the result. In order to ensure that human health is adequately protected, the public health assessment incorporates conservative (unlikely to underestimate risk) approaches and uncertainty factors. Therefore, the actual risk associated with the exposure routes considered in this assessment is unlikely to be larger, but may be lower than that predicted.

TABLE 5-1

Estimated Risks from Soil Ingestion at the Crop Duster Site

Scenario 1 - Short-term Exposure, Current Conditions

Non-carcinogenic hazard

	Chronic Daily Intake	CDI/RfD Ratio
	Children	
DDT:	5.5 x 10^{-5}mg/kg/day	0.1
	Adult Workers	
DDT:	3.0 x 10^{-6}mg/kg/day	0.006

Carcinogenic risk

	Chronic Daily Intake	Cancer Risk
	Children	
DDT:	7.8 x 10^{-7}mg/kg/day	3 x 10^{-7}
Toxaphene:	4.4 x 10^{-6}mg/kg/day	5 x 10^{-6}
	Total	5 x 10^{-6}
	Adult Workers	
DDT:	4.3 x 10^{-8}mg/kg/day	1 x 10^{-8}
Toxaphene:	2.3 x 10^{-7}mg/kg/day	3 x 10^{-7}
	Total	3 x 10^{-7}

TABLE 5-2

Estimated Risk from Airborne DDT and Toxaphene at the Crop Duster Site

Scenario 1 -Short-term Exposure, Current Conditions

Non-carcinogenic hazard

Compound	Chronic Daily Intake	CDI/RfD
DDT:	1.9×10^{-7}mg/kg/day	<0.001

Carcinogenic risk

Compound	Chronic Daily Intake	Cancer Risk
DDT:	2.9×10^{-9}mg/kg/day	1×10^{-9}
Toxaphene:	1.6×10^{-8}mg/kg/day	2×10^{-8}
	Total	2×10^{-8}

TABLE 5-3

Estimated Risks from Soil Ingestion at the Crop Duster Site

Scenario 2- Long-term Exposure, Current Conditions

Non-carcinogenic hazard

	Chronic Daily Intake	CDI/RfD Ratio
	Children	
DDT:	3.3×10^{-5}mg/kg/day	0.07
	Adult Workers	
DDT:	3.0×10^{-6}mg/kg/day	0.003

Carcinogenic risk

	Chronic Daily Intake	Cancer Risk
	Children	
DDT:	2.2×10^{-6}mg/kg/day	7×10^{-7}
Toxaphene:	1.3×10^{-5}mg/kg/day	2×10^{-5}
	Total	2×10^{-5}
	Adult Workers	
DDT:	4.3×10^{-7}mg/kg/day	1×10^{-7}
Toxaphene:	2.3×10^{-6}mg/kg/day	3×10^{-6}
	Total	3×10^{-6}

TABLE 5-4
Estimated Risk from Airborne DDT and Toxaphene at the Crop Duster Site
Scenario 2- Long-term Exposure, Current Conditions

Non-carcinogenic hazard

Compound	Chronic Daily Intake	CDI/RfD
DDT:	1.9×10^{-7} mg/kg/day	<0.001

Carcinogenic risk

Compound	Chronic Daily Intake	Cancer Risk
DDT:	8.7×10^{-8} mg/kg/day	3×10^{-8}
Toxaphene:	4.8×10^{-7} mg/kg/day	6×10^{-7}
	Total	7×10^{-7}

TABLE 5-5:
Estimated Risks from Soil Ingestion at the Crop Duster Site

Scenario 4 - Residential Exposure, Potential Future Use

Non-carcinogenic hazard

Children

	Chronic Daily Intake	CDI/RfD Ratio
DDT:	1.9×10^{-4} mg/kg/day	0.4

Carcinogenic risk

Resident Between the Ages of 1 - 30 Years

	Chronic Daily Intake	Cancer Risk
	Children	
DDT:	2.7×10^{-5} mg/kg/day	9×10^{-6}
Toxaphene:	1.4×10^{-4} mg/kg/day	2×10^{-4}
	Total	2×10^{-4}

TABLE 5-6

Estimated Risk from Airborne DDT and Toxaphene at the Crop Duster Site

Scenario 4 - Residential Exposure, Potential Future Use

Non-carcinogenic hazard

Compound	Chronic Daily Intake	CDI/RfD
DDT:	1.9×10^{-7}mg/kg/day	<0.001

Carcinogenic risk

Compound	Chronic Daily Intake	Cancer Risk
DDT:	2.9×10^{-9}mg/kg/day	1×10^{-9}
Toxaphene:	1.6×10^{-8}mg/kg/day	2×10^{-8}
	Total	2×10^{-8}

6.0 CONCLUSIONS

Based on the results of this health risk assessment, ΣDDT at the airport site appears unlikely to cause systemic (non-carcinogenic) effects at the site. An approved Reference Dose was not available for toxaphene. However, NAS (1977) reported an ADI of 0.001 mg/kg/day for toxaphene. Considering the relative difference between this value and the RfD for DDT, and further considering the difference in concentrations at the site, toxaphene also appears unlikely to pose a hazard, except possibly under the residential use scenario, and the cancer risks for this scenario are also above EPA target criteria.

Inhalation exposure to fugitive dust containing organochlorine pesticides poses a health risk below the 10^{-6} target cancer risk level used by the State and EPA. Consequently, air emissions from the site do not appear likely to pose a significant risk. On the other hand, direct contact and subsequent ingestion of site soils was estimated to pose an excess cancer risk above the EPA target risk goal of 10^{-6} under most of the scenarios considered in the assessment. However, cancer risks associated with contact with site soils were generally within EPA's target risk range of 10^{-4} - 10^{-6}. Toxaphene was the chemical responsible for most of this elevated cancer risk. Toxaphene is not uniformly distributed across the site but tends to be concentrated in an area adjacent to the former airport taxiway and closed mix system area. Frequent contact with this area would likely pose an excess cancer risk somewhat above levels estimated in this report. However, the positive side is that focusing remedial activities in this area would probably substantially mitigate the problem without the cost of attempting to remediate the entire site.

Use of the property for commercial office space, particularly if the construction was designed to ensure capping of the areas containing elevated levels of pesticides, would be unlikely to lead to significant exposure or risks. Other exposure scenarios that involve similar containment of the pesticide-containing soils would also be unlikely to pose significant health risks.

In order to ensure that public health is adequately protected, conservative (unlikely to underestimate risk) assumptions were used in deriving both the exposure estimate and the toxicity values. Because of the use of these conservative (although not necessarily worst case) assumptions, it is likely that actual risks are considerably lower than risks estimated in this report. For example, much of the airport runway only contains DDT and at levels that do not pose a non-cancer health risk. If DDT acts by an epigenetic mechanism these low levels may pose no actual health hazard.

7.0 REFERENCES

Agency for Toxic Substances and Disease Registry (ATSDR) 1989. Toxicological Profile for p,p'-DDT, p,p'-DDE, p,p'-DDD. U.S. Department of Health and Human Services (DHHS). December 1989.

Agency for Toxic Substances and Disease Registry (ATSDR) 1990. Toxicological Profile for Toxaphene. U.S. Department of Health and Human Services (DHHS). December 1990.

Anderson, E., Browne, N., Duletsky, S., Ramig, J., and Warn, T., Development of Statistical Distributions or Ranges of Standard Factors Used in Exposure Assessments. Prepared for the Office of Health and Environmental Assessment, U.S. Environmental Protection Agency, Washington, D.C. March 1985. OHEA-E-161. (1985)

Carey, A.E., Gowen, J.A., Tai, H., Mitchell, W.G., and Wiersma, G.B. 1979. Pesticide Residue Levels in Soils and Crops from 37 States, 1972 -- National Soils Monitoring Program (IV). Pesticides Monitoring Journal 12: 209-229.

Environmental Protection Agency (EPA) 1979. Water Related Environmental Fate of 129 Priority Pollutants. Washington, D.C. December 1979. EPA 440/4-79-029

Environmental Protection Agency (EPA). 1986. Guidelines for the Health Risk Assessment of Chemical Mixtures. Fed. Reg. 51: 34013-34023

Environmental Protection Agency (EPA). 1988. Superfund Exposure Assessment Manual. Office of Remedial Response, USEPA, Washington, D.C., EPA 540/1-88-001 April (1988)

Environmental Protection Agency (EPA). Exposure Factors Handbook. Office of Health and Environmental Assessment, Washington, D.C. July 1989. EPA 600/8-89/043, July (1989)

Environmental Protection Agency (EPA). Risk Assessment Guidance for Superfund: Human Health Evaluation Manual, Part A (Interim Final). Office of Solid Waste and Emergency Response, Washington, D.C. OSWER Directive 9285.701A, July (1989)

Environmental Protection Agency (EPA), Human Health Evaluation Manual, Supplemental Guidance: "Standard Default Exposure Factors" Office of Solid Waste and Emergency Response, Washington, D.C. March 1991. OSWER Directive 9285. 6-03, (1991)

Environmental Protection Agency (EPA), Risk Assessment Guidance for Superfund: Human Health Evaluation Manual, Part B: Development of Risk-Based Preliminary Remediation Goals (Interim). Office of Solid Waste and Emergency Response, Washington, D.C. July 1989. OSWER Directive 9285.7-01B (1991)

Environmental Protection Agency (EPA), Integrated Risk Information System (IRIS). Chemical-Specific Reference Doses and Cancer Potency Factors and EPA Toxicology Background Documents. Office of Health and Environmental Assessment, Environmental Criteria and Assessment Office, Cincinnati, OH (1993)

LaGoy, P. K., and Hopkins, L., Developing Site-Specific Cleanup Levels: Practical Considerations, *Remediation*, 1:113-121 (1991)

Sedman, R., The Development of Applied Action Levels for Soil Contact: A Scenario for the Exposure of Humans to Soil in a Residential Setting, Final Draft. Prepared by Toxic Substances Control Division (1987)

APPENDIX B - CASE STUDY 2

Risk Assessment-Based Remediation Goals for a Metal Plating Company

This case study incorporates the procedures described in Chapter 8 for developing and justifying site-specific cleanup levels for chemicals in soils and groundwater. A semi-fictitious metal plating company (L&M Metalworks Plating Company) in the Midwestern U.S. is used as the example site. The site is semi-fictitious in that the chemical concentrations, settings, etc. are drawn from past work but no such single property actually exists.

1.0 INTRODUCTION

This document develops and justifies site-specific cleanup levels for chemicals in soils and groundwater at the L&M Metalworks Plating Company in the Midwestern United States. The remediation goals developed in this report follow guidance presented in EPA's (1991) Risk Assessment Guidance for Superfund: Volume 1 - Human Health Evaluation Manual (Part B): Development of Risk-Based Preliminary Remediation Goals (PRGs). The PRGs are non-site-specific values designed to be used as a screening tool to focus remedial efforts on key areas of potential concern. Chemical concentrations at the site are compared with the PRGs to establish the need for further evaluation or remediation. Based on the results of this comparison, site-specific remediation goals are developed for chemicals with on-site concentrations exceeding the PRGs.

Section 2 of the report contains background information on the site and presents the key chemicals of concern. The next section presents information on the environmental behavior and toxicity of key chemicals. As part of the toxicological assessment, health-based criteria are identified. Section 4 contains a discussion of exposure pathways for the site and also includes the calculation of PRGs. Cleanup goals established by various state agencies are also discussed in this section and final site-specific cleanup levels are presented. The comparison of the risk-based remediation goals with on-site chemical concentrations is presented in Section 5. Also as part of this section, a preliminary assessment of potential remedial options, and a consideration of risk issues related to selection of these options is described. The final section, Section 6, provides the conclusions of the risk assessment.

2.0 SITE CHARACTERIZATION

This section of the risk assessment contains a brief description of the history, location, topography, geology, and climate of the site. A summary of the nature and extent of chemicals at the site is also included (ed. note: the term contamination can be used in place of chemicals if appropriate and acceptable to the client).

2.1 Site Background

The L&M Metalworks Plating site is comprised of two properties, the actual plating operations facility and an adjacent lot where sampling indicated additional chemical contamination. The site is located a quarter mile south of the Missouri River and is bordered by Moody Road and a highway right of way to the north, by a utility company to the south, and by local residences to the east and west (see Figure 1). The surrounding area is comprised of light and heavy industry, commercial operations, and local residences.

Plating operations began at the site in 1968 and ceased in 1985. Previously, the site had been residential. During the plating operations, the company performed zinc, nickel, cadmium, chrome, and copper plating, using a series of plating solutions, acid washes, and soap rinses.

Process chemicals used on the site include sodium cyanide, zinc cyanide, copper cyanide, cadmium, chromic acid, hydrochloric acid, and sodium hydroxide. Trichloroethylene (TCE) and 1,1,1-trichloroethane (TCA) were used for degreasing metal parts prior to plating.

Houses immediately east of the site are inhabited but the houses to the west are empty and in disrepair. Children live in several of the houses to the east. During a site visit, a number of children were seen in the area around the site. Although the site is fenced, it appears that children could climb the fence and gain access to the site without too much difficulty.

2.2 General Site Characteristics

The surface topography of the site is flat, with less than 1 percent slope across the site. Currently, berms constructed around the site prevent surface water runoff from the facility. The L & M property is partially covered with a concrete slab and exposed soils are generally covered with grasses. Therefore, the potential for wind erosion is low. The soils beneath the site consist of clay and silty clays, with some gravel. The depth to groundwater is approximately 15 feet. A number of shallow wells, which may be used for irrigation or industrial purposes at the present time, are located in the vicinity of the site. None are known to be used for domestic water supply. The average annual rainfall in the area is 38 inches/year, with a 1-year, maximum 24-hr rainfall of 4 inches.

2.3 Chemical Characterization of the Site

Numerous studies have been conducted to characterize the nature and extent of contamination at the site. Initial sampling at the site was conducted by the State Environmental Control Agency in 1979. Visible contamination was noted in the back portion of the facility and sampling in this area revealed elevated concentrations of cadmium, chromium, copper, zinc, cyanide, and chlorinated solvents. An excavation program was initiated and soil in the visibly contaminated area was removed to a depth of one foot. Subsequent samples were collected by the State in 1985 and 1986 from on-site containers, from discolored site surface areas, and from surface soils. These samples also indicated the presence of heavy metals and cyanide.

In June 1986, the State's contractor collected samples from the vacant lot east of the property and from the adjacent utility company's land to the south. Sampling at these locations was performed because of the presence of discolored soils, lack of vegetation, and the potential for human exposure. The vacant lot east of the site was unfenced and children had been observed playing there. Runoff from the rear of the former operations facility onto the utility company property was also suspected. Samples from these locations revealed the presence of heavy metals, cyanide, and halogenated VOCs.

Complete information on sampling and analysis at the site is presented in a Remedial Investigation report. Sampling results pertinent to developing risk based cleanup goals are presented in Tables 1 and 2. Table 1 presents the detection frequency, limits of detection (LODs), arithmetic mean and maximum concentrations of metals, cyanide, and halogenated solvents in on-site soil samples, and sample locations for the maximum concentrations detected. Table 2 presents this information for three background locations and also presents national background data from Shacklette and Boerngen (1984). Sample identification numbers corresponding to the locations at which the maximum concentrations were found for on-site and background samples are also presented in the tables; sample locations are presented in Figure 2. Chemicals selected as chemicals of concern based on their presence at levels above background concentrations are presented in bold in Table 1. Only these chemicals are considered in this report.

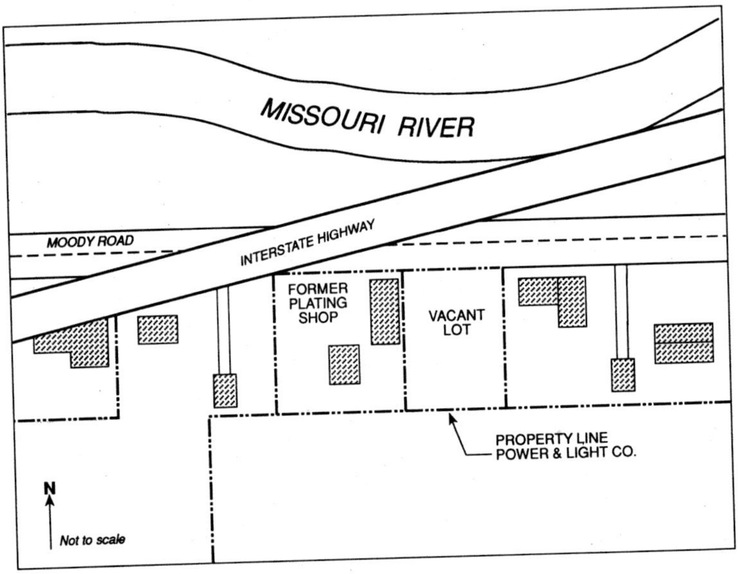

Figure 1
SITE LOCATION MAP
L&M Metalworks Site

TABLE 1: Concentrations of Chemicals in Soil at the Site

Chemical	No. of Detections/ No. of Samples	LOD[a] (mg/kg)	Arithmetic[b] Mean (mg/kg)	Maximum Conc. (mg/kg)	Sample Location of Maximum[c]
Arsenic	4/25	5,50	7	9	15
Beryllium	0/25	0.75,1.0,7.5	ND[d]	ND	_
Cadmium	**18/25**	**1,10**	**62**	**530**	**8**
Chromium[e]	**19/25**	**50**	**190**	**2230**	**8**
Copper	**12/25**	**50**	**260**	**3040**	**8**
Lead	24/25	50	110	460	1
Mercury	7/25	0.2,2	0.8	4	C
Nickel	12/25	100	47	130	B
Silver	0/25	5,50	ND	ND	-
Zinc	**21/25**	**100**	**620**	**4200**	**8**
Cyanide	**22/25**	**1**	**170**	**2530**	**8**
1,1,1-TCA	**4/25**	**0.01**	**5**	**11**	**6**
TCE	**4/25**	**0.01**	**12**	**38**	**6**

[a]Limits of detection can vary from one sample to the next due to interferences, sample dilution, and other analytical considerations.

[b]Arithmetic mean of detected values only

[c]See Figure 2 for sample locations.

[d]Not detected

[e]All chromium was assumed to be CRVI based on limited sampling.

TABLE 2: Concentrations of Chemicals in Background Soil Samples from the Vicinity of the Site

Chemical	No. of Detections/ No. of Samples	LODs[a] (mg/kg)	Arithmetic[b] Mean (mg/kg)	Maximum Conc. (mg/kg)	Sample Location of Maximum
Arsenic	1/3	5	9	9	BKG-16
Beryllium	0/3	0.75, 1.0	ND	ND	-
Cadmium	2/3	1	1.4	1.5	BKG-18
Chromium	3/3	-	43	53	BKG-18
Copper	3/3	-	15	21	BKG-16
Lead	3/3	-	130	260	BKG-16
Mercury	1/3	0.2	0.2	0.2	BKG-16
Nickel	3/3	-	30	59	BKG-18
Silver	0/3	5	ND	ND	-
Zinc	3/3	-	130	225	BKG-18
Cyanide	1/3	1	0.6	0.6	BKG-18

[a]Limits of detection can vary among samples. All reported LODs are shown in this column.
[b]Arithmetic mean of detected values only. Background samples BKG-16, BKG-17, and BKG-18 were used to calculate the arithmetic mean concentration.
[c]Not detected

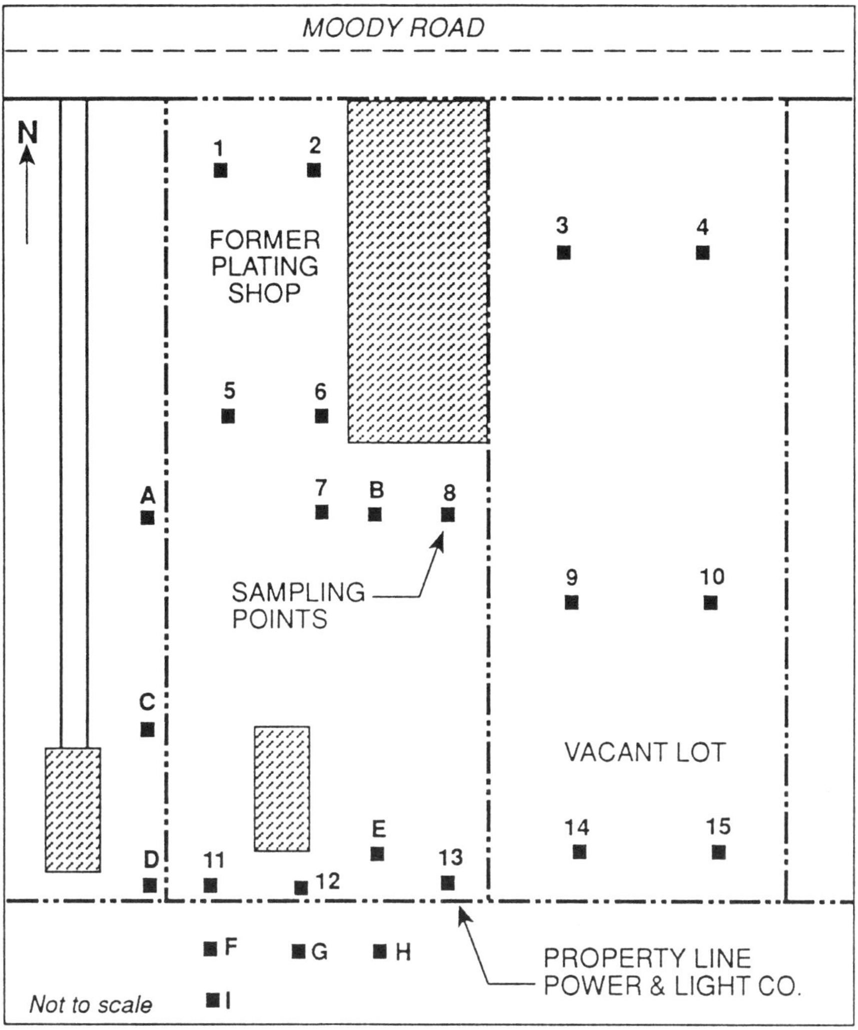

Figure 2

SAMPLING LOCATIONS

L&M Metalworks Site

3.0 CHARACTERIZATION OF KEY CHEMICALS

3.1 Environmental Transport and Fate

The environmental behavior of a particular chemical is dependent on the physical and chemical properties of the compound, the environmental transformation processes affecting them, and properties of the media in which it is located. Because the potential for exposure is highly dependent on the fate and transport of the chemicals of concern, these characteristics are summarized for chemicals detected at the site. Chemicals of concern at the site include several metals, halogenated volatile organic compounds (VOCs) used as degreasing agents, and the anion cyanide. The properties of these chemicals are summarized below.

Metals. Metals are generally not considered to be very mobile in the environment and are likely to remain bound to site soils. However, the behavior of a particular metal is highly dependent on its form in the environment and infiltrating precipitation may leach the more soluble chemicals from soil and transport them into the groundwater. Soil parameters that must be considered are clay and metal oxide content, fraction of organic matter, pH, and oxidation-reduction potential. Chemical partitioning between soil and water can be expressed by a soil-water distribution coefficient (K_d). K_ds may underestimate migration potential because they are based on chemical adsorption and desorption reaching equilibrium. As a rough estimate of relative mobilities, K_ds for the inorganic chemicals of concern can be compared. For example, cadmium has a low K_d of 6.5 whereas lead, with a K_d of 900, tends to be much less mobile. However, mobility can be enhanced by the presence of other chemicals such as naturally occurring humic materials. In general, metals are not considered to be particularly mobile in the environment.

Metals adsorbed to surface soil particles may migrate into the air as a component of windborne fugitive dust. Release of large quantities of fugitive dust is not expected to occur because most of the area is covered by buildings or grass. Therefore, migration of metals from the soil surface and subsequent exposure via inhalation is considered unlikely.

Cyanide. The term cyanide can refer to any material containing the CN group; these may include simple cyanides, metallocyanide complexes, and organic compounds (nitriles) (EPA 1979). The simple cyanides are comprised of HCN and those salts formed with Group I and Group II metals, such as NaCN and $Ca(CN)_2$. These rapidly dissociate in water, leaving CN- to hydrolyze or to react with trace metals. Iron, gold, cadmium, copper, nickel, silver, and zinc have been found to form simple metallocyanides (Fuller 1977, 1978). Simple metallocyanides have solubilities ranging from essentially insoluble (copper and nickel) to highly soluble (barium and potassium). Complex cyanides are formed when cyanide bonds covalently to a metal, forming an anionic (negatively charged) complex. Cadmium and zinc complexes dissociate in water while the stability of copper and nickel cyanides depends on pH. Iron and cobalt cyanide complexes are very stable (EPA 1979, EPA 1985).

The mobility of cyanide in five soils was studied by Alesii and Fuller (1976). Their results indicated that while salts dissociated to release ions which were mobile in soils, pH, metal oxide and clay content in the soils significantly influenced the mobility. CN^- is retained in soils with high concentrations of manganese and iron oxides. $Fe(CN)_6^{-3}$ is retained in soils with high clay content and at low pH.

Degradation of cyanide can occur both by chemical oxidation and by anaerobic or aerobic biodegradation, with the ultimate breakdown products being CO_2 and NH_3. In the aerobic microbial environment, cyanide first undergoes enzymatic conversion to cyanate, which is then converted to carbon dioxide and ammonia (Van Zyl 1985). The rate at which such biodegradation occurs has not been determined, but is probably highly variable and will be dependent on many factors including the types of microbial colonies present as well as the nature and concentration of the waste.

The fate of cyanide in soils will depend on the stability of any metallocyanide species and on the biological activity of the soil. Formation of stable complexes with iron and other metals can lead to cyanide persisting in soils. Ultimately, hydrolysis of the cyanide species and subsequent biodegradation, together with volatilization of HCN from surface soils, will reduce the soil concentration.

Volatile Organic Compounds. The volatile organic chemicals (VOCs) generally have relatively low organic carbon partition coefficients (K_{oc}s) indicating they are not likely to adsorb to the soil organic matter.

Most of these chemicals have solubilities in the hundreds-of-milligrams-per-liter range or greater (1,000 mg/liter or ppm for trichloroethylene) and are expected to be fairly mobile in the aqueous phase (i.e. once they become dissolved in groundwater).

The halogenated VOCs are rather volatile and are likely to be released from uncovered vadose zone soils into the ambient air. The potential for compounds to volatilize from exposed subsurface soils can be predicted by their vapor pressure and/or Henry's Law constants, with compounds having high vapor pressures or Henry's Law constants generally volatilizing readily.

Laboratory studies and field observations have shown that chlorinated aliphatic hydrocarbons can undergo both biological and abiological (physicochemical) transformations in soil and groundwater under the right conditions. However, the compounds tend to be fairly resistent to degradation, and the products of degradation may be equally or more toxic than the parent compounds. For example, a degradation product of TCE (oral cancer slope factor of $(0.01\ mg/kg/day)^{-1}$) is vinyl chloride (oral cancer slope factor of $(1.9\ mg/kg/day)^{-1}$), which is one hundred times more potent.

3.2 Toxicity and Health Criteria

Brief descriptions of the toxic effects of the key site chemicals are presented below. Toxicity criteria (cancer slope factors (CSF) for carcinogens and reference doses (RfDs) for noncarcinogens) are presented in Table 3.

Cadmium. Cadmium has not been shown to be carcinogenic when administered orally. Oral exposure can lead to an accumulation of cadmium in the kidneys with subsequent dysfunction (EPA 1985). The EPA (1993) calculated oral reference doses of 5×10^{-4} mg/kg/day for cadmium in water and 1×10^{-3} mg/kg/day for cadmium in food. The EPA (1992) promulgated a maximum contaminant level (MCL) of 5 ug/liter for cadmium in drinking water.

The EPA has classified cadmium in Group B1 -- Probable Human Carcinogen (by inhalation only), based on the weight of the evidence for carcinogenicity in mice and occupationally exposed workers (EPA 1985).

The EPA's Carcinogen Assessment Group (CAG) calculated a "unit risk" for inhalation exposure of 1.8×10^{-3} $(ug/m^3)^{-1}$ (equivalent to a cancer potency factor of 6.1 $(mg/kg/day)^{-1}$) (EPA 1993).

Chromium/Chromium VI. The toxicity of chromium is dependent to a great degree on the chemical species. Certain chromium (VI) compounds are considered to be carcinogenic by inhalation (but not by ingestion) and chromium (III) compounds are not considered to be very toxic by any exposure route. Studies in humans and animals using a variety of routes of administration have indicated that soluble chromium (VI) and chromium (III) compounds can produce kidney and liver damage following acute exposure to relatively high doses. From the evidence available from both human case reports and animal studies, it is uncertain whether the kidneys and liver may be target organs following chronic exposure. The EPA has verified an oral reference dose (RfD) for chromium (III) of 1 mg/kg/day and has calculated a RfD of 5×10^{-3} mg/kg/day for chromium (VI). These values are based on studies in which no adverse health effects were produced in rats exposed to chromium compounds in the diet or drinking water. The EPA promulgated an MCL of 0.1 mg/liter for total chromium.

TABLE 3

Oral Toxicity Criteria for Key Chemicals

Chemical	Reference Dose mg/kg/day	Cancer Slope Factor $(mg/kg/day)^{-1}$
Cadmium	5×10^{-4}	
Chromium VI	5×10^{-3}	
Copper	4×10^{-2}	
Zinc	3×10^{-1}	
Cyanide	2×10^{-2}	
1,1,1-TCA	9×10^{-2}	
TCE	6×10^{-3}	1.1×10^{-2}

Source: EPA 1993, Smith 1993

The EPA (1984) determined a cancer potency factor for lifetime inhalation exposure to chromium (VI) of 41 $(mg/kg/day)^{-1}$ based on occupational studies (Mancuso 1975). This is equivalent to a unit risk of $1.2x10^{-2}$ $(ug/m^3)^{-1}$ assuming a 70-kg person inhales 20 m^3 air/day. EPA (1993) has classified chromium (VI) in Group A--Human Carcinogen (by inhalation), based on the weight of the evidence for carcinogenicity.

Copper. Copper is not a potent toxic agent to humans (EPA 1993). The EPA Office of Drinking Water has established an MCL for copper of 1.3 mg/liter based on the occurrence of gastrointestinal effects in man (EPA 1992). EPA (1993) has determined an oral RfD for copper of $3.7x10^{-2}$ mg/kg/day.

Zinc. Zinc is not highly toxic to humans and is, in fact, an essential element to human nutrition (EPA 1985). Zinc salts of strong acids are astringent and corrosive. An oral reference dose of 0.3 mg/kg/day was derived by EPA (1993) based on the occurrence of anemia and reduced blood copper in some patients receiving therapeutic doses of approximately 150 mg/day zinc for extended periods of time. A secondary MCL of 5 mg/liter has been developed for zinc but a primary MCL has not been developed.

Cyanide. Hydrogen cyanide and its simple salts, such as sodium cyanide, are acutely toxic to almost all forms of life by all exposure routes (EPA 1980). Cyanide has not been tested for carcinogenicity in animals. The EPA (1993) derived an oral RfD of 0.02 mg/kg/day for free cyanide and hydrogen cyanide based on a study in which rats were fed cyanide-containing diets for 104 weeks with no adverse effects observed (EPA 1986).

1,1,1-Trichloroethane. Exposure to high levels of 1,1,1-trichloroethane (TCA) has been associated with central nervous-system depression and cardiovascular effects including premature ventricular contractions and arrhythmias. Fatty changes have been reported in rodent livers following inhalation exposure. Two chronic carcinogenicity bioassays have been conducted on 1,1,1-trichloroethane; suggestive evidence of an increased incidence of hepatocellular tumors in female mice was reported in one study but both bioassays were flawed. EPA currently classifies TCA in Group D -- Not Classified. The EPA (1993) calculated an RfD for TCA of $9x10^{-2}$ mg/kg/day. An MCL of 0.2 mg/liter has been promulgated (EPA 1992).

Trichloroethylene. Trichloroethylene (TCE) is a central nervous system depressant following acute or chronic exposure, by both ingestion and inhalation. Industrial use of TCE may also result in dermatitis from exposure to vapors of concentrated solvent. In carcinogenicity studies on mice, an increased incidence of hepatocellular carcinomas was reported following oral administration of TCE (EPA 1993). The EPA (1993) has derived cancer slope factors of 1.1×10^{-2} $(mg/kg/day)^{-1}$ for oral exposure and 6×10^{-3} $(mg/kg/day)^{-1}$ for inhalation exposure. The oral cancer slope factor is currently undergoing review but will be used in this assessment pending development of an alternate value. TCE has been classified by EPA in Group B2--Probable Human Carcinogen. The EPA's Office of Drinking Water has promulgated an MCL of 5 μg/liter for TCE in drinking water (EPA 1992).

4.0 CLEANUP LEVEL DEVELOPMENT

The development of cleanup levels includes both the simple calculation of risk-based goals and a consideration of site-specific factors that will affect the potential for exposure. This section presents information on potential exposure pathways at the site, contains the calculation of preliminary remediation goals for these pathways, and finally, lists cleanup criteria established by various state and federal agencies.

4.1 Exposure Pathways

Exposure to the metals and cyanide at the L & M Metalworks site is most likely to occur as a result of direct contact with the on-site soil. The majority of the metals are persistent, not very mobile in soil, and will probably not migrate from the site in substantial quantities. Chromium VI and cyanide are relatively mobile, but migration from the site will tend to be limited by chemical transformation processes and by the clay content of the soils at the site. The migration of TCE and 1,1,1-TCA will also be limited to some degree but these chemicals are likely to be the most mobile constituents at the site. Groundwater in the area currently is not used for domestic purposes but domestic use will be considered as a maximum exposure scenario.

The site is currently fenced and access is likely to be limited. Primarily for this reason, the site does not currently appear to pose a public health risk to area residents. Children or trespassers who may climb the fence and engage in activities resulting in direct soil contact may receive some exposure and if the fence were removed, the potential for exposure would also increase.

4.2 Preliminary Remediation Goal Development

The EPA (1991) has established procedures for determining Preliminary Remediation Goals (PRGs) for chemicals at hazardous waste sites. These PRGs use maximum exposure scenarios and conservative assumptions so that it is unlikely that contact with the chemical-affected media of concern at the PRG levels would pose a significant health risk under almost any condition. The PRGs generally do not consider site-specific factors, and depending on site characteristics and use, concentrations substantially greater than the PRGs may also not pose health risks. The PRGs are designed primarily to be used as a screening tool; if all levels of a chemical are below the PRGs, no further work is necessary. However, if the PRGs are exceeded, further analysis of the cleanup criteria including evaluation of site-specific factors is warranted and remediation may be needed. A factor that needs to be considered in the assessment is the potential for multiple chemical exposure. If exposure to a large number of chemicals is possible, a value equal to the PRG divided by a safety factor of five or ten can be used in the screening analysis. For this site, a factor of the PRG/5 will be used for metals and cyanide to evaluate the risks for direct contact exposure. The PRGs alone will be used to evaluate the threat to groundwater as only the two VOCs are considered for this pathway. PRGs were developed only for soils as groundwater samples were not collected. However, the PRGs for the halogenated VOCs were developed to protect groundwater as a water supply resource.

The reduced equation for calculating PRGs for direct contact exposure to noncarcinogenic chemicals in residential soils is:

$$PRG = 2.7 \times 10^5 (RfD_o)$$

Where RfD_o is the oral reference dose for the chemical.

For direct contact exposure to carcinogens in residential soils, the reduced equation is:

$$PRG = 0.64/SF_o$$

Where SF_o is the cancer slope factor for the chemical.

For nonresidential (industrial/commercial) land use, the reduced equations consider both direct ingestion and inhalation exposure. However, for persistent chemicals, volatilization is generally minimal except immediately after application, and even for more volatile chemicals, their presence at the site for long periods of time probably substantially reduces the potential for volatilization. Eliminating the inhalation portion of the equation, the formula for **noncarcinogens** reduces to:

$$PRG = 2 \times 10^6 (RfD_o)$$

For carcinogens the equation is:

$$PRG = 5.8/SF_o$$

The RfDs or cancer slope factors and the weight of the evidence for carcinogenicity for carcinogens are presented in Table 3 for the chemicals of concern. Soil PRGs calculated for both residential and commercial/industrial exposure are presented in Table 4 along with risk-based concentrations (RBCs) developed by EPA Region III. The most stringent criteria (carcinogenic vs. noncarcinogenic) were used in establishing the values.

Volatile compounds such as the halogenated solvents are of greatest concern due to their potential to impact groundwater. Therefore PRGs and RBCs based on direct contact exposure are inadequate. Site-specific modelling is commonly used to establish allowable concentrations that if left in soils would not produce groundwater concentrations exceeding the groundwater PRGs. As a screening tool, assuming site groundwater is not currently being used as a water supply source, a simplified approach can be used to develop a soil PRG protective of groundwater. This value is based on an assumed dilution and attenuation by soil of a factor of 100. The PRG_{GW} is established by multiplying the MCL (in mg/liter) by the dilution/attenuation factor (DAF) of 100 and by 1 liter/kg to convert from volumetric measure to mass to arrive at a soil PRG in mg/kg. These values are also presented in Table 4.

TABLE 4

Preliminary Remediation Goals and EPA Region III Risk-Based Concentrations (mg/kg)

	PRGs			Region III RBCs	
Constituent	Residential	Commercial	Subsurface	Residential	Commercial
Cadmium	135	2,000	-	**39**	510
Chromium	1,350	10,000	-	**390**	5,100
Copper	10,000	10,000	-	**2,900**	10,000
Zinc	10,000	10,000	-	**10,000**	10,000
Cyanide	5,400	10,000	-	**1,600**	10,000
1,1,1-TCA	10,000	10,000	**20**	7,000	10,000
Trichlorothene	60	520	**0.5**	110	260

TABLE 5

Soil Criteria Developed by Other States (mg/kg)

	State Regulatory Agency				
		New Jersey[b]		Michigan[c]	
Constituent	Washington[a]	Surface	Subsurface	Direct	GW Protect
Cadmium	2	1	100	100	0.08
Chromium	100	NA	NA	1,000	2
Copper	250	600	600	10,000	20
Zinc	NA	1500	1500	50,000	20
Cyanide	NA	1100	NA	2,000	2
1,1,1-TCA	20	210	50	2,000	4
Trichlorothene	0.5	23	1	40	0.06

[a]Health-based criteria established by Washington State for either groundwater protection or direct contact protection. The state of Tennessee has adopted the Washington State Soil Criteria.

[b]New Jersey values for surface (direct contact exposure) or subsurface (groundwater protection). The lower of the two values is generally used as the criteria.

[c]Michigan values based on direct contact exposure or 20 times the groundwater value in mg/liter.

4.3 Available Cleanup Criteria

Soil remediation levels designed to adequately protect humans from any potential health effects have been developed by the states of New Jersey, Washington, and Michigan, and are presented in Table 5. A comparison of the site-specific PRGs with the criteria developed by the regulatory agencies indicates that the values are fairly consistent as long as the route of exposure used as the basis for the values are consistent. The PRGs listed in bold on Table 4 will be used in this evaluation. These PRGs and the regulatory risk-based criteria were developed with minimal consideration of site-specific factors that may influence risk, and consequently are best used to screen site locations to evaluate the potential for adverse health effects.

5.0 RISK CHARACTERIZATION

In this section, the PRGs from Table 4 are compared with the reported soil concentrations to determine if on-site concentrations exceed the preliminary target criteria. If all on-site concentrations are below target criteria, no further action is necessary at the site. However, if on-site concentrations do exceed these levels, some remediation may be necessary and further evaluation to consider the impact of this remediation on the toxicity, mobility, and risks posed by the chemicals is required.

5.1 Comparison of Site Concentrations with Remediation Goals

The PRGs from Table 4 were compared with the reported maximum soil concentrations (from Table 1) to determine if on-site concentrations exceed the preliminary target criteria. Based on this comparison, only 1,1,1-TCA was below the PRG. Copper, zinc, and cyanide all exceeded the PRG only at a single location (sample point 8). Chromium exceeded the PRG at 4 locations, all near the southern edge of the building. TCE exceeded the criteria for groundwater protection at all 4 locations where it was detected and cadmium was detected at levels exceeding the soil criteria in 8 of the 25 samples.

5.2 Site-Specific Goals

As noted by EPA in the PRG guidance document, "The recommended approach for developing remediation goals is to identify PRGs at scoping, modify them as needed at the end of the RI or during the FS based on site-specific information from the baseline risk assessment, and ultimately select remediation levels in the Record of Decision (ROD)." Concentrations of most of the chemicals detected on site exceed the PRGs in at least one location. Consequently, final remediation goals that were less stringent than the PRGs but that would still be suitably protective of human health considering the intended use of the property were developed. The most likely future use of the property is for commercial/industrial purposes. A comparison of commercial/industrial criteria (from Table 4) with the maximum on-site concentration for the chemicals suggests that this use would not pose a significant risk for direct contact exposure. TCE would still pose a threat to groundwater but groundwater is not used for domestic purposes currently, and such use is unlikely in the future. Groundwater from the site flows towards the Missouri River and the chance of a domestic well being installed between the site and the river is small. Consequently, if future use is industrial/commercial, remediation may be unnecessary.

A key component of a risk assessment and of the PRG development process is an identification of the level of uncertainty associated with the approach. Uncertainty factors in this evaluation that lead to overestimating potential risks include the use of health-protective assumptions and approaches in developing preliminary remediation goals, the assumption that soil is uncovered, and the assumption that frequent, regular use of the site will occur. Factors that may lead to underestimating potential risks include not accounting for multiple chemical exposure and not accounting for exposure through pathways considered unlikely to occur (i.e., vegetable gardening). On balance, the use of conservative (health protective; unlikely to underestimate risk) assumptions and approaches in deriving these preliminary remediation goals suggests that it is unlikely that chemicals remaining at the site at these levels would pose an actual hazard.

If no remedial action is taken at the L & M Metalworks site and site use is not restricted (either practically, by use for industrial purposes, or by deed restrictions), use of the site in the future may differ from its current use. Construction of residential dwellings on the site at some point in the future could pose a risk to residents due to exposure to lead, cadmium, chromium and nickel. Use of the site for vegetable gardens might also pose a greater risk than evaluated in this assessment.

6.0 CONCLUSIONS

Based on a comparison with residential-use PRGs, some remediation would be necessary prior to allowing unrestricted access to the property. Based on a comparison with commercial/industrial PRGs, direct contact with site soils is unlikely to pose a risk and remediation may not be necessary. TCE is present in soils at levels that may impact groundwater use for domestic purposes. The presence of TCE also affects remediation options, as halogenated compounds are prohibited from land disposal without treatment. Some of the options available to the property owner include:

No Action

Unacceptable without access/deed restrictions; acceptable for industrial use. Cost negligible except in terms of limits on use of the land.

Off-Site Disposal

Acceptable and would allow unrestricted use of land; costs for disposing of relatively small amount of soil (500 yd^3) with no metals exceeding TCLP levels would be \$100/yd^3 for a total cost of \$50,000. However, cyanide from metal plating, a RCRA listed waste and TCE in soils would require treatment at a substantial cost and probably bring overall removal costs to the \$500/ton range or around \$400,000 for the total project.

Off-Site Disposal after On-Site Treatment

Acceptable and would allow unrestricted use of land; costs for disposing of treated RCRA-listed waste with metal levels below TCLP levels would be around \$100,000. Treatment to remove most TCE through soil vapor extraction would cost around \$50,000 for a total cost of \$150,000

On-Site Containment

Acceptable but might not allow unrestricted future use. In addition, long-term groundwater monitoring would be required and considering that future use would not be unrestricted, this approach would cost more than and be no more effective than the no-action option.

NOTE: After considering all possible options, the landowner can make an informed decision about possible action at the site. An additional option might be to perform a more detailed evaluation of risk issues using a Monte Carlo-type approach. Depending on the state agency involved, a somewhat more detailed risk assessment, possibly including further information on the potential bioavailability of the chemicals may be sufficient to allow unrestricted use (the difference between the target levels and the average soil concentration is fairly small). Alternatively, it may be possible based on a more detailed risk assessment showing only limited risks, to convince the agency to allow unrestricted access after removal of a small hotspot area at substantially reduced costs. Predicting the outcome of negotiations on risk issues is essentially impossible without knowing the agency involved. However, over time as risk issues become better understood, it should become easier to make convincing arguments.

7.0 REFERENCES

Alesi, B.A., and Fuller, W.H., *The mobility of three cyanide forms in soils*. In: Fuller, W.H., (ed.), Residual Management by Land Disposal, Proceedings of the Hazardous Waste Research Symposium. EPA 600/7-76-015, NTIS PB 256 768, July (1976)

Environmental Protection Agency (EPA). 1979. The Water-Related Environmental Fate of the 129 Priority Pollutants. Prepared for USEPA, Office of Water Planning and Standards, Washington, D.C. December 1979. EPA 440/4-79-029A and B

Environmental Protection Agency (EPA). 1985. Chemical, Physical, and Biological Properties of Compounds Present at Hazardous Waste Sites. Prepared by Associates for Office of Waste Program Enforcement, Washington, D.C., September (1985)

Environmental Protection Agency (EPA). 1989. Risk Assessment Guidance for Superfund: Human Health Evaluation Manual, Part A, Interim Final. Office of Solid Waste and Emergency Response,

Washington, D.C., OSWER Directive 9285.701A, July (1989)

Environmental Protection Agency (EPA), Risk Assessment Guidance for Superfund: Human Health Evaluation Manual, Part B: Development of Risk-Based Preliminary Remediation Goals (Interim). Office of Solid Waste and Emergency Response, Washington, D.C., July 1989, OSWER Directive 9285.7-01B (1991)

Environmental Protection Agency (EPA) 1992. Drinking Water Regulations and Health Advisories. Office of Water, U.S. EPA Washington, D.C. December 1992

Environmental Protection Agency (EPA), Integrated Risk Information System (IRIS). Chemical-Specific Reference Doses and Cancer Potency Factors and EPA Toxicology Background Documents. Office of Health and Environmental Assessment, Environmental Criteria and Assessment Office, Cincinnati, OH (1993)

Frasco, B. 1993. Soil Cleanup Criteria (New Jersey). Site Remediation News, Volume 5(1). April 1993.

LaGoy, P.K., Why Washington State has the Best Cleanup Standards Now, *Remediation*, Summer (1992) 331-340

Michigan Department of Natural Resources 1992. MERA Operational Memorandum #8, Revision 1 - Type B Criteria, Rules 299.5709, 299-5711(2), 299.5711(5) and 299.5713. From A.J. Howard, Chief, Environmental Response Division, to Environmental Response Division Staff, March 16, 1992.

Shacklette, H.T., and Boergnen, J.G., Element Concentrations in Soils and Other Surficial Materials of the Coterminous United States. U.S. Geological Survey Professional Paper 1270, U.S. Government Printing Office, Washington, D.C. (1984)

Smith, R.L. 1993. Risk-Based Concentration Table, Third Quarter 1993. U.S. Environmental Protection Agency, Region III, Philadelphia, Pennsylvania.

REFERENCES

1. Environmental Protection Agency (EPA). Risk Assessment Guidance for Superfund: Volume 1 -- Human Health Evaluation Manual (Part A). Fact Sheet, OSWER, USEPA Washington, D.C., April 1990. OSWER 9285.7-01/FS (1990)

2. Environmental Protection Agency (EPA). Superfund 30-Day Task Force Report: Accelerating Superfund Cleanups and Evaluating Risk at Superfund Sites. OSWER, July 19, (1991)

3. Environmental Protection Agency (EPA), Risk Assessment Guidance for Superfund: Human Health Evaluation Manual, Part A, Interim Final. Office of Solid Waste and Emergency Response, Washington, DC, OSWER Directive 9285.701A, July (1989)

4. Burmaster, D.E. and Lehr, J.H., It's Time to Make Risk Assessment a Science, *Ground Water Monitoring Report*, Summer pp 5-15 (1991)

5. Paustenbach, D.J., Introduction: A Primer for Conducting Human or Environmental Health Risk Assessment, in: *The Risk Assessment of Environmental Hazards: A Textbook of Case Studies.* (Paustenbach, D.J., ed.) John Wiley & Sons, New York. (1989)

6. Environmental Protection Agency (EPA), Superfund Public Health Evaluation Manual. Office of Emergency and Remedial Response, Washington, D.C. (OSWER Directive 9285.4-1) EPA 540/1-86/060 October (1986)

7. Environmental Protection Agency (EPA), Exposure Factors Handbook. Office of Health and Environmental Assessment, Washington, DC, EPA 600/8-89/043, July (1989)

8. Environmental Protection Agency (EPA), Human Health Evaluation Manual, Supplemental Guidance: "Standard Default Exposure Factors." Office of Solid Waste and Emergency Response, Washington, D.C., March 1991. OSWER Directive 9285.6-03 (1991)

9. Environmental Protection Agency (EPA), Risk Assessment Guidance for Superfund: Human Health Evaluation Manual, Part B: Development of Risk-Based Preliminary Remediation Goals (Interim). Office of Solid Waste and Emergency Response, Washington, D.C. July 1989. OSWER Directive 9285.7-01B (1991)

10. Environmental Protection Agency (EPA), Superfund Exposure Assessment Manual. Office of Remedial Response, USEPA, Washington, DC, EPA 540/1-88-001, April (1988)

11. LaGoy, P.K., Nisbet, I.C.T. and Schulz, C.O., Endangerment Assessment for the Smuggler Mountain Site in Pitkin County, Colorado: A Case Study, in: *The Risk Assessment of Environmental Hazards: A Textbook of Case Studies* (D.J. Paustenbach, ed.), pp.505-525, John Wiley and Sons, New York, NY (1989)

12. Environmental Protection Agency (EPA), Integrated Risk Information System (IRIS). Chemical-Specific Reference Doses and Cancer Potency Factors and EPA Toxicology Background Documents. Office of Health and Environmental Assessment, Environmental Criteria and Assessment Office, Cincinnati, OH, (1993)

13. Environmental Protection Agency (EPA), Guidelines for Carcinogen Risk Assessment. Fed. Reg. 51:33992-34003 (1986)

14. MacKay, D.M., Roberts, P.V., and Cherry, J.A., Transport of organic contaminants in groundwater. Environ. Sci. Technol. 19(5) 384-392 (1985)

15. Vogel, T.M. and McCarty, P.L., Biotransformation of Tetrachloroethylene to Trichloroethylene, Dichloroethylene, Vinyl

Chloride, and Carbon Dioxide under Methanogenic Conditions. *Appl. Environ. Microbiol.* 49:1080-1083 (1985)

16. Lyman, W.J., Reehl, W.F., and Rosenblatt, D.H., *Handbook of Chemical Property Estimation Methods: Environmental Behavior of Organic Compounds.* McGraw-Hill Book Co., New York. (1982)

17. Baes, C.F., Sharp, R.D., Sjoreen, A.L., and Shor, R.W., A Review and Analysis of Parameters for Assessing Transport of Environmentally-Released Radionuclides through Agriculture. Prepared by the Oak Ridge National Laboratory operated by Martin Marietta Energy Systems, Inc. for the U.S. Department of Energy. ORNL-5786 (1984)

18. Bouwer, E.J. and McCarty, P.L., Transformation of 1- and 2-Carbon Halogenated Aliphatic Organic Compounds under Methanogenic Conditions. *Appl. Environ. Microbiol.* 45:1286-1294 (1983)

19. Bouwer, E.J., Rittmann, B.E., and McCarty, P.L., Anaerobic Degradation of Halogenated 1 and 2-Carbon Organic Compounds. *Environ. Sci. Technol.* 15:596-599 (1981)

20. LaGoy, P. K., Significant Figures in Risk assessment: More Caution is Warranted (Letter to the Editor), *Risk Analysis*, 9:431-432 (1989)

21. Schneiderman, M.A., Decoufle, P. and Brown, C.C., Thresholds for Environmental Cancer: Biologic and Statistical Considerations, *Annals of the New York Academy of Sciences.* 329:92-130 (1979)

22. Zapp, J.A., Industrial Toxicology: Retrospect and Prospect, in: *Patty's Industrial Hygiene and Toxicology.* Vol. 1: General Principles. 3rd. ed. (Clayton, G.D. and Clayton, F.E., eds.) Wiley-Interscience Publication, John Wiley and Sons, New York, NY (1978)

23. Carlborg, F.W., Observable Tumours, Risk Assessment and Experimental Design, *Fd Chem. Toxic.* 22(1):69-73 (1984)

24. Environmental Protection Agency (EPA), Health Effects Assessment Summary Tables: Annual: FY-1992. Office of Emergency and Remedial Response, Washington, DC, OERR 9200.6-303-(92-1),

March (1992)

25. Timbrell, J.A., *Principles of Biochemical Toxicology.* Taylor and Francis LTD., London (1982)

26. Doull, J., Klaussen, D.C. and Amdur, M.O., *Casarett and Doull's Toxicology: The Basic Science of Poisons.* 2nd ed. Macmillan Publishing Co., New York (1986)

27. Environmental Protection Agency (EPA), Dermal Exposure Assessment: Principles and Applications (Interim Report) Office of Research and Development, U.S. EPA January 1992, EPA/600/8-91/011B (1992)

28. Young, S.S. and Gries, C.L., Exploration of the Negative Correlation between Proliferative Hepatocellular Lesions and Lymphoma in Rats and Mice-Establishment and Implications, *Fundamental and Applied Toxicology* 4:632-640 (1984)

29. Kimbrough, R.D., Mitchell, F.L., and Houk, V.N., Trichloroethylene: An Update, *J. of Toxicology and Environmental Health* 15:369-383 (1985)

30. Marafante, E., Bertolero, F., Edel, J., Pietra, R., and Sabbioni, E., Intracellular Interaction and Biotransformation of Arsenite in Rats and Rabbits. Sci. Total Env. 24: 27-39. (1982)

31. Haseman, J.K., Huff, J.E., Rao, G.N., Arnold, J.E., Boorman, G.A., and McConnell, E.E., Neoplasms Observed in Untreated and Corn Oil Gavage Control Groups of F344/N Rats and (C57BL/6N X C3H/HeN)F1 (B6C3F1) Mice, *J. Natl. Cancer Inst.* 75(5):975-984 (1985)

32. Ross, M.H., Lustbader, E.D., and Bras, G., Body Weight, Dietary Practices, and Tumor Susceptibility in the Rat, *J. Natl. Cancer Inst.* 71(5):1041-1046 (1983)

33. Ross, M.H., Lustbader, E.D. and Bras, G., Dietary Practices of Early Life and Age at Death of Rats with Tumors, *J. Natl. Cancer Inst.* 71(5):947-954 (1983)

34. Lehman, A.J., *Appraisal of the Safety of Chemicals in Foods, Drugs, and Cosmetics*, Association of Food and Drug Officials in the U.S. (1959)

35. Ecobichon, D.J., *The Basis of Toxicity Testing*, CRC Press, Inc., Boca Raton, FL (1992)

36. Dourson, M.L., New Approaches in the Derivation of Acceptable Daily Intake (ADI), *Comments in Toxicology* 1:35-48 (1986)

37. Dourson, M.L. and Stara, J.F., Regulatory History and Experimental Support of Uncertainty (Safety) Factors, *Regulatory Toxicology and Pharmacology* 3:224-238 (1983)

38. Dacre, J.C., Rosenblatt, D.H., and Cogley, D.R., Preliminary Pollutant Limit Values for Human Health Effects, *Environ. Sci. Technol.* 14(7):778-784 (1980)

39. Davidson, I.W.F., Parker, J.C., and Beliles, R.P., Biological Basis for Extrapolation across Mammalian Species, *Regulatory Toxicology and Pharmacology* 6:211-237 (1986)

40. Goldstein, B.D., The Problem with the Margin of Safety: Toward the Concept of Protection, *Risk Analysis* 10:7-10 (1990)

41. Layton, D.W., Mallon, B.J., Rosenblatt, D.H., and Small, M.J., Deriving Allowable Daily Intakes for Systemic Toxicants Lacking Chronic Toxicity Data, *Regulatory Toxicology and Pharmacology* 7:96-112 (1987)

42. Sielken, R.L., Jr., Cancer Dose-Response Extrapolations, *Environ. Sci. Technol.* 21(11):1033-1039 (1987)

43. Anderson, E.L. and the Carcinogen Assessment Group of the U.S. Environmental Protection Agency, Quantitative Approaches in Use to Assess Cancer Risk, *Risk Analysis* 3(4):277-295 (1983)

44. Menzel, D.B., Physiological Pharmacokinetic Modeling, *Environ. Sci. Technol.* 21(10):944-950 (1987)

45. Moolgavkar, S.H., Carcinogenesis Modeling: From Molecular Biology to Epidemiology, *Ann. Rev. Public Health* 7:151-169 (1986)

46. Moolgavkar, S.H. and Knudson, A.G., Jr., Mutation and Cancer: A Model for Human Carcinogenesis, *J. Natl. Cancer Inst.* 66(6):1037-1052 (1981)

47. Environmental Protection Agency (EPA), Drinking Water Regulations and Health Advisories. Office of Water, U.S. EPA, Washington, D.C. December (1992)

48. American Conference of Governmental Industrial Hygienists (ACGIH), Threshold Limit Values and Biological Exposure Indices for 1992-1993. Cincinnati, Ohio. (1992)

49. American Conference of Governmental Industrial Hygienists (ACGIH), Documentation of the Threshold Limit Values and Biological Exposure Indices., 5th Ed., Cincinnati, OH (1986)

50. Graham, J.D., Green, L.A., and Roberts, M.J., *In Search of Safety: Chemicals and Cancer Risk,* Harvard University Press, Cambridge, MA (1991)

51. Griesemer, R.A. and Cueto, C., Jr., Toward a Classification Scheme for Degrees of Experimental Evidence for the Carcinogenicity of Chemicals for Animals, in: *Molecular and Cellular Aspects of Carcinogen Screening Tests* (R. Montesana, H. Bartsch and L. Tomatis, eds.) IARC Scientific Publications 27:259-281 (1980)

52. International Agency for Research on Cancer (IARC), *IARC Monographics on the Evaluation of the Carcinogenic Risk of Chemicals to Humans, Vol. 39: Some Chemicals Used in Plastics and Elastomers.* World Health Organization, Lyon, France. (1986)

53. Ames, B.N., Dietary Carcinogens and Anticarcinogens, Oxygen Radicals and Degenerative Diseases, *Science* 221:1256-1264 (1983)

54. Hammond, E.C., Smoking in Relation to the Death Rates of One Million Men and Women, in: *Epidemiological study of Cancer and Other Chronic Diseases* (W. Haenszel, ed.) NCI Monograph 19:127-204 (1966)

55. Environmental Protection Agency (EPA), Chemical, Physical and Biological Properties of Compounds Present at Hazardous Waste Sites. Prepared by Clement Associates for Office of Waste Program Enforcement, Washington, DC, September (1985)

56. Manson, J.M., Teratogens, in: *Casarett and Doull's Toxicology--The Basic Science of Poisons*. (Doull, J., Klaassen, C., Amdur, M., eds.) 2nd ed. Macmillan Publishing Co., New York (1986)

57. Nisbet, I.C.T., and Karch, N.J., *Chemical Hazards to Human Reproduction*, Noyes Data Corporation, Park Ridge, New Jersey (1983)

58. Ballantyne, B., Evaluation of Hazards from Mixtures of Chemicals in the Occupational Environment, *J. of Occupational Medicine* 27(2):85-94 (1985)

59. Mahaffey, K.R. and Fowler, B.A., Effects of Concurrent Administration of Lead, Cadmium, and Arsenic in the Rat, *Environmental Health Perspectives* 19:165-171 (1977)

60. Thilly, W.G., Longwell, J., and Andon, B.M., General Approach to the Biological Analysis of Complex Mixtures, *Environmental Health Perspectives* 48:129-136 (1983)

61. Environmental Protection Agency (EPA), Guidelines for the Health Risk Assessment of Chemical Mixtures. Fed. Reg. 51: 34013-34023 (1986)

62. Bellinger, D., Leviton, A., Waternaux, C., Needleman, H., and Rabinowitz, M., Longitudinal Analyses of Prenatal and Postnatal Lead Exposure and Early Cognitive Development, *New Eng. J. Med.* 316:1037-1043 (1987)

63. Environmental Protection Agency (EPA), Review of the National

Ambient Air Quality Standards for Lead: Exposure Analysis Methodology and Validation. OAQPS Staff Report. Office of Air Quality Planning and Standards, Research Triangle Park, NC, EPA 450/2-89-011, June (1989)

64. Chaney, R.L., Sterrett, S.B., and Mielke, H.W., The Potential for Heavy Metal Exposure from Urban Gardens and Soils in: *Proc. Symp. Heavy Metals in Urban Gardens* (J.R. Preer, ed.) pp. 37-84, Univ. Dist. Columbia Extension Service, Washington, DC (1984)

65. Centers for Disease Control (CDC), Preventing Lead Poisoning in Young Children. Public Health Service, Chronic Disease Division, Atlanta, GA, July (1985)

66. Environmental Protection Agency (EPA), Air Quality Criteria for Lead, Draft Final. Environmental Criteria and Assessment Office, Research Triangle Park, NC, EPA 600/8-83-028aF, June (1986)

67. Shacklette, H.T. and Boerngen, J.G., Element Concentrations in Soils and Other Surficial Materials of the Coterminous United States. U.S. Geological Survey Professional Paper 1270, U.S. Government Printing Office, Washington, DC (1984)

68. Travis, C.C. and Hattemer-Frey,H.A., A Perspective on Dioxin Emissions from Municipal Solid Waste Incinerators, *Risk Analysis* 9:91-98 (1989)

69. North Atlantic Treaty Organization (NATO), Pilot Study on International Information Exchange on Dioxins and Related Compounds. International Toxicity Equivalency Factor (I-TEF) Method of Risk Assessment for Complex Mixtures of Dioxins and Related Compounds. Committee on the Challenges of Modern Society. Report No. 176, August (1988)

70. Centers for Disease Control (CDC), Letter from R. Kimbrough, Medical Officer, CDC, to D. Wagoner, Director, Waste Management Division, EPA Region VII, RE: Review of "Remedial Alternatives for a Dioxin-Contaminated Site," January 13, (1987)

71. Centers for Disease Control (CDC), Letter from D. Forney, Public

Health Advisor, to S. Margolis, Environmental Health Manager, RE: Review of "Proposed Workplan to Address the Dioxin Contamination Problem in Newark, New Jersey," November 20, (1984)

72. Kimbrough, R.D., Falk, H., and Stehr, P., Health implications of 2,3,7,8-TCDD contamination of residential soil, *J. Toxicol. Environ. Health* 14: 47-93 (1984)

73. LaGoy, P.K., Bohrer, R.L., and Halvorsen, F.H., The Development of Cleanup Criteria for an Acutely Toxic Pesticide at a Contaminated Industrial Facility, *Am. Ind. Hyg. Assoc. J.* 53:298-302 (1992)

74. Beck, B.D., Connoly, R.B., Dourson, M.L., Guth, D., Hattis, D., Kimmel, C. and Lewis, S.C., Improvements in Quantitative Non-Cancer Risk Assessment, *Fund. Appl. Toxicol.* 20:1-14 (1993)

75. Habicht, F.H., II Guidance on Risk Characterization for Risk Managers and Risk Assessors (Memorandum from Deputy Administrator to Assistant administrators and Regional Administrators) US EPA, Washington, D.C. February 26 (1992)

76. Crouch, E.A.C., Pilkinton, M.B.G., and Zemba, S.G., The Effects of "Hot Spots" and Upper Confidence Limits in Estimating Exposure Concentrations in Soils in: *Superfund Risk Assessment in Soil Contamination Studies, ASTM STP 1158* (K.B. Hoddinott and G.D. Knowles, eds) American Society for Testing and Materials, Philadelphia, PA (1992)

77. Schaum J. 1984. Risk Analysis of TCDD-contaminated soil. Office of Health and Environmental Assessment, Environmental Protection Agency, Washington, D.C. November 1984. EPA 600/8-84-031

78. Hawley, J.K., Assessment of Health Risk from Exposure to Contaminated Soil, *Risk Analysis* 5:289-302 (1985)

79. Sedman, R., The Development of Applied Action Levels for Soil Contact: A Scenario for the Exposure of Humans to Soil in a Residential Setting, Final Draft. Prepared by Toxic Substances Control Division (1987)

80. Anderson, E., Browne, N., Duletsky, S., Ramig, J., and Warn, T., Development of Statistical Distributions or Ranges of Standard Factors Used in Exposure Assessments. Prepared for the Office of Health and Environmental Assessment, U.S. Environmental Protection Agency, Washington, D.C. March 1985. OHEA-E-161. (1985)

81. Ott, W.R., Total Human Exposure: Basic Concepts, EPA Field Studies, and Future Research Needs, *J. of the Air & Waste Management Association* 40(7):966-975 (1990)

82. Wallace, L.A., The Total Exposure Assessment Methodology (TEAM) Study, Project Summary, USEPA, Office of Acid Deposition, Environmental Monitoring and Quality Assurance, Washington DC (1987)

83. Wallace, L.A., Pellizzari, E.D., Hartwell, T.D., Sparacino, C.M., Sheldon, L.S., and Zelon, H., Personal Exposures, Indoor-Outdoor Relationships, and Breath Levels of Toxic Air Pollutants Measured for 355 Persons in New Jersey, *Atmospheric Environment* 19(10):1651-1661 (1985)

84. Environmental Protection Agency (EPA), Development of Emission Factors for Fugitive Dust Sources, Final Report. Prepared by Midwest Research Institute for USEPA Research Triangle Park, NC, EPA 450/3-74-037, June (1974)

85. Brown, H.S., Bishop, D.R., and Rowan, C.A., The Role of Skin Absorption as a Route of Exposure for Volatile Organic Compounds (VOCs) in Drinking Water, *Am. J. of Public Health* 74(5):479-484 (1984)

86. Foster, S.A. and Chrostowski, P.C., Inhalation Exposures to Volatile Organic Contaminants in the Shower. Presented at the 80th Annual APCA Meeting. New York, NY, June 21-26 (1987)

87. Little, J.C., Applying the Two-Resistance Theory to Contaminant Volatilization in Showers, *Environ. Sci. Technol.* 26(7):1341-1349 (1992)

88. McKone, T.E., Human Exposure to Volatile Organic Compounds in

Household Tap Water: The Indoor Inhalation Pathway, *Environ. Sci. Technol.* 21(12):1194-1200 (1987)

89. LaGoy, P.K., Estimated Soil Ingestion Rates for Use in Risk Assessment, *Risk Analysis* 7:355-359 (1987)

90. Poiger, H. and Schlatter, C., Influence of Solvents and Adsorbents on Dermal and Intestinal Absorption of TCDD, *Fd Cosmet. Toxicol.* 4 18:477-481 (1980)

91. Umbrett, T.H., Hesse, E.J. and Gallo, M.A., Bioavailability of Dioxin in Soil from a 2,4,5-T Manufacturing Site, *Science* 232:497-499 (1986)

92. Feldmann, R.J. and Maibach, H.I., Percutaneous Penetration of some Pesticides and Herbicides in Man, *Toxicology and Applied Pharmacology* 28:126-132 (1974)

93. Howd, R.A., Hsieh, D.P.H., Reed, W.A., and Christopher, J.P., Dermal Absorption of Toluene and Xylene from Soil, Presentation #19 at the NorCal SOT meeting, San Francisco, CA (1989)

94. McKone, T.E. and Ryan, P.B., Human Exposures to Chemicals through Food Chains: An Uncertainty Analysis, *Environ. Sci. Technol.* 23:1154-1163 (1989)

95. Environmental Protection Agency (EPA), Special Report on Ingested Inorganic Arsenic - Skin Cancer; Nutritional Essentiality. Risk Assessment Forum, USEPA, Washington, D.C. EPA 625/3-87/013 (1988)

96. Shea, D., Developing National Sediment Quality Criteria, *Environ. Sci. Technol.* 22(11):1256-1261 (1988)

97. Environmental Protection Agency (EPA), Ambient Water Quality Criteria for Copper. Office of Water Regulations and Standards, Criteria and Standards Division, Washington, DC, EPA 440/5-80-036, October (1980)

98. Burmaster, D.E., Murray, D.M., and Anderson, P.D., Distributions for

the Average Daily Consumption Rates of Self-Caught and Store-Bought Fish by Ontario Adults Who Ate Fish from the Great Lakes in 1988, (submitted for publication; May 1993)

99. Maugh, T.H., II, Biological Markers for Chemical Exposure, *Science* 215:643-647 (1982)

100. Wibowo, A.A.E., Herber, R.F.M., Das, H.A., Roeleveld, N., and Zielhuis, R.L., Levels of Metals in Hair of Young Children as an Indicator of Environmental Pollution, *Environmental Research* 40:346-356 (1986)

101. Carlisle, J.C. and Wade, M.J., Predicting Blood Lead Concentrations from Environmental Concentrations, *Regulatory Toxicology and Pharmacology* 16:280-289 (1992)

102. Chrostowski, P.C., A Comparison of the Integrated Uptake/Biokinetic Model to Traditional Exposure Assessment Approaches, in: *Superfund Risk Assessment in Soil Contamination Studies,* ASTM STP 1158 (K.B. Hoddinott and G.D. Knowles, eds.) American Society for Testing and Materials, Philadelphia, PA (1992)

103. Environmental Protection Agency (EPA), National Primary Drinking Water Regulations-Synthetic Organic Chemicals and Inorganic Chemicals; Monitoring for Unregulated Contaminants; National Primary Drinking Water Regulations Implementation; National Secondary Drinking Water Regulations. Fed. Reg. 56:3526-3597, January 30 (1991)

104. LaGoy, P. K. and Hopkins, L., Developing Site-Specific Cleanup Levels: Practical Considerations, *Remediation* 1:113-121 (1991)

105. Finkel, A.M., Is Risk Assessment Really Too Conservative?: Revising the Revisionists, *Columbia Journal of Environmental Law* 14:427-467 (1989)

106. Whipple, C., Nonpessimistic Risk Assessment and de Minimus Risk as Risk Management Tools, in: *The Risk Assessment of Environmental Hazards: A Textbook of Case Studies.* (Paustenbach, D.J., ed.) John Wiley & Sons, New York. (1989)

107. Green, T. and Prout, M.S., Species Differences in Response to Trichloroethylene, II. Biotransformation in Rats and Mice, *Toxicology and Applied Pharmacology* 79:401-411 (1985)

108. Prout, M.S., Provan, W.M. and Green, T., Species Differences in Response to Trichloroethylene, I. Pharmacokinetics in Rats and Mice, *Toxicology and Applied Pharmacology* 79:389-400 (1985)

109. Thompson, K.M. and Burmaster, D.E., Parametric Distributions for Soil Ingestion by Children, *Risk Analysis* 11:339-342 (1991)

110. Thompson, K.M., Burmaster, D.E., and Crouch, E.A.C., Monte Carlo Techniques for Quantitative Uncertainty Analysis in Public Health Risk Assessments. *Risk Analysis* 12: 99-104 (1992)

111. D'Appolonia, E., Monitored Decisions. *J. Geotechnical Engineering* 116: 4-34. (1990)

112. Brown, S.M., Lincoln, D.R., and Wallace, W.A., Application of the Observational Method to Remediation of Hazardous Waste Sites, Unpublished Report, CH2M HILL, Bellevue, WA April (1989)

113. Maugh, T.H., II, The Dump That Wasn't There, *Science* 215:645 (1982)

114. Morgan, M.G., Risk Analysis and Management, *Scientific American* July:32-41 (1993)

115. Slovic, P., Informing and Educating the Public About Risk, *Risk Analysis* 6(4):403-415 (1986)

116. Covello, V.T. and Allen, F.W., Seven Cardinal Rules of Risk Communication, EPA pamphlet, Washington, DC (1988)

117. Rutgers, 10 Ways to Lose Trust and Credibility, Environmental Communication Research Program pamphlet

118. Tversky, A. and Kahneman, D., The Framing of Decisions and the Psychology of Choice, *Science* 211:453-458 (1981)

119. Covello, V.T., Sandman, P.M. and Slovic, P., Risk Communication, *Risk Statistics, and Risk Comparisons: A Manual for Plant Managers,* The Chemical Manufacturers Association, Washington, D.C. (1988)

120. Covello, V.T., Communicating Right-To-Know Information on Chemical Risks, *Environ. Sci. Technol.* 23(12):1444-1449 (1989)

121. LaGoy, P.K. Why Washington State has the Best Cleanup Standards Now, *Remediation* Summer (1992) 331 - 340.

122. Michigan Department of Natural Resources (MDNR) Environmental Contamination Response Activity Administrative Rules for 1982 PA 307 as amended (1990), updated as: Memorandum from A.J. Howard, Chief, Environmental Response Division to Environmental Response Division Staff; MERA Operational Memorandum #8, Revision 1 - Type B Criteria Rules March 16, (1992)

123. New Jersey Department of Environmental Protection and Energy (DEPE) Proposed New Rule - Cleanup Standards for Contaminated Sites, February 1992, updated as: Frasco, B., Soil Cleanup Criteria, *Site Remediation News* 5:1-16 (1993)

124. Park, K.S., Sims, R.C., Dupont, R.R., Doucette, W.J., and Matthews, J.E., Fate of PAH Compounds in Two Soil Types: Influence of Volatilization, Abiotic Loss and Biological Activity, *Environmental Toxicology and Chemistry* 9:187-195 (1990)

125. Cordle, F., Locke, R. and Springer, J., Risk Assessment in a Federal Regulatory Agency: An Assessment of Risk Associated with the Human Consumption of Some Species of Fish Contaminated with Polychlorinated Biphenyls (PCBs), *Environmental Health Perspectives* 45:171-182 (1982)

126. Smith, R.L., Risk-Based Concentration Table, Third Quarter 1993, Memorandum from R.L. Smith, Senior Toxicologist to RBC Table Mailing List, U.S. EPA, Region III, Philadelphia, PA July (1993)

127. Shapiro, J., Radiation Protection: *A Guide for Scientists and Physicians*, Second Edition, Harvard University Press, Cambridge, MA (1981)

128. National Conference on Radiation Protection and Measurement (NCRP), *Evaluation of Occupational and Environmental Exposures to Radon and Radon Daughters in the United States*. NCRP Report No. 78, Bethesda, Maryland (1984)

129. Hileman, B., Indoor Air Pollution, *Environ. Sci. Technol.* 17(10):469A-472A (1983)

130. Hobbs, C.H. and McClellan, R.O., Radiation and Radioactive Materials. in: *Casarett and Doull's Toxicology--The Basic Science of Poisons*. (Doull, J., Klaassen, C., Amdur, M., eds.) 2nd ed. Macmillan Publishing Co., New York (1986)

131. Environmental Protection Agency (EPA), *Background Information Document--Proposed Standards for Radionuclides*. U.S. Office of Radiation Programs, Washington, D.C. March 1983. EPA 520/1-83-001 (1983)

132. Committee on the Biological Effects of Ionizing Radiation (BEIR III). *The Effects on Populations of Exposure to Low Levels of Ionizing Radiation:* National Academy Press, Washington, D.C. (1980)

133. Environmental Protection Agency (EPA), Risk Assessment Guidance for Superfund: Environmental Evaluation Manual, Interim Final. Office of Emergency and Remedial Response. EPA 540/1-89/001A (1989)

134. Environmental Protection Agency (EPA), Framework for Ecological Risk Assessment. Risk Assessment Forum, USEPA, Washington, D.C. February 1992 EPA 630/R-92/001, (1992)

135. Warren-Hicks, W., Parkhurst, B.R., and Baker, S.S., eds. 1989. Ecological Assessments of Hazardous Waste Sites: A Field and Laboratory Reference Document. Prepared by Kilkelly Environmental Associates, Raleigh, N.C. for the Environmental Protection Agency Office of Research and Development. EPA600/3-89/013 (1989)

136. Schafer, E.W., Jr., Bowles, W.A., Jr. and Hurlbut, J., The Acute Oral Toxicity, Repellency, and Hazard Potential of 998 Chemicals to One or More Species of Wild and Domestic Birds, *Arch. Environm.*

Contam. Toxicol. 12:355-382 (1983)

137. Plafkin, J.L., Barbour, M.T., Porter, K.D., Gross, S.K., and Hughes, R.M., Rapid Bioassessment Protocols for Use in Streams and Rivers: Benthic Macroinvertebrates and Fish, U.S.EPA Assessment and Watershed Protection Division, Washington, D.C. (1989)

138. U.S. Army Corps of Engineers Wetlands Delineation Manual. Waterways Experimental Station, Vicksburg, MS. Technical Report Y-87-1 (1987)

139. Christian, J.J., Love Canal's Unhealthy Voles, *Natural History* 10:8-16 (1983)

140. Richkind, K.E. and Hacker, A.D., Responses of Natural Wildlife Populations to Air Pollution, *J. of Toxicology and Environmental Health* 6:1-10 (1980)

141. Port, G.R. and Thompson, J.R., Outbreaks of Insect Herbivores on Plants Along Motorways in the United Kingdom, *J. of Applied Ecolgy* 17:649-656 (1980)

142. Nisbet, I.C.T., Ecological Magnification, *Technology Review* March/April (1975)

143. Nisbet, I.C.T. and Reynolds, L.M., Organochlorine Residues in Common Terms and Associated Estuarine Organisms, Massachusetts, USA, *Marine Environmental Research* 11:33-66 (1984)

144. Birge, W.J., Black, J.A., and Ramsey, B.A., The Reproductive Toxicology of Aquatic Contaminants, *Hazard Assessment of Chemicals: Current Developments* 1:59-115 (1981)

INDEX